HUMAN ANATOMY AND PHYSIOLOGY

TRYPTOPHAN: DIETARY SOURCES, FUNCTIONS AND HEALTH BENEFITS

HUMAN ANATOMY AND PHYSIOLOGY

Additional books in this series can be found on Nova's website under the Series tab.

Additional E-books in this series can be found on Nova's website under the E-book tab.

HUMAN ANATOMY AND PHYSIOLOGY

TRYPTOPHAN: DIETARY SOURCES, FUNCTIONS AND HEALTH BENEFITS

BLAKE L. WHITLEY
AND
SARAH H. THORNTON
EDITORS

Nova Biomedical Books
New York

For permission to use material from this book please contact us:
Telephone 631-231-7269; Fax 631-231-8175
Web Site: http://www.novapublishers.com

Library of Congress Cataloging-in-Publication Data

ISBN 978-1-62100-445-5

Library of Congress Control Number: 2011035652

Published by Nova Science Publishers, Inc. † *New York*

Contents

Preface

This book presents topical research in the study of the dietary sources, functions and health benefits related to tryptophan. Topics discussed include molecular imprinting for chiral separation of tryptophan; tryptophan ingestion improves the synthesis of serotonin and melatonin and may be related with delay of some conditions of aging; the role of tryptophan residues in antimicrobial activity and membrane interactions; the health benefits from tryptophan supplementation in humans and the metabolism of tryptophan and evaluation of tryptophan supplmentation on fish larval growth and frequency of skeletal deformities.

Chapter I - Molecular imprinting is an emerging methodology for the creation of selective recognition sites in synthetic polymers. This technique entails the polymerization of functional monomers in the presence of an important molecule (template). Recent studies have shown that the polymers obtained exhibit a surprisingly high degree of stereo- and regiospecific selectivity, making the commercial use of such tailor-made separation materials in the area of chiral separation of bioactive molecules a realistic possibility. This chapter summarizes the present state of the art of molecular imprinting to generate tailor-made chiral stationary phases (CSPs) for the chiral separation of tryptophan and provides an overview of the major factors involved in the manufacturing process that are crucial to the chromatographic performance of the phases.

Chapter II - The essential amino acid tryptophan is the precursor in the biosynthesis of the indoleamines serotonin and melatonin. Brain serotonin is synthesized in two reactions beginning with the hydroxylation of tryptophan to 5-hydroxytryptophan which is then decarboxylated to serotonin.

As the hydroxylation of tryptophan is the rate-limiting step in the synthesis of serotonin, tryptophan hydroxylase (TPH) determines the effective concentration of serotonin *in vivo*. TPH is now known to exist in two isoforms: TPH-1 is mainly expressed in the pineal gland and in gut enterochromaffin cells, and TPH-2 is preferentially expressed in the brain.

The TPH activity increases after tryptophan ingestion which shows that the rate-limiting enzyme is far from being saturated in normal conditions, and confirms that the synthesis of serotonin and melatonin can be modulated by tryptophan ingestion.

The administration of L-tryptophan during the light time increases the brain synthesis and metabolism of serotonin. At night, tryptophan's administration led to a smaller increase in the synthesis of serotonin than by day, although the turnover remainunchanged, implying that, in the dark phase, serotonin is used as a substrate for melatonin synthesis. As a consequence, the amount available of this hormone is dependent first on an adequate dietary supply of tryptophan, and second on the balance between serotonin's use as a neurotransmitter and its availability as precursor for melatonin synthesis which is deeply dependent on the environmental light.

As age advances, the nocturnal production of melatonin decreases. It can be partly explained by a decrease of serum/plasma tryptophan concentration in humans related with aging and associated with an enhanced indoleamine (2,3)-dioxygenase activity, which degrades tryptophan to form kynurenine derivatives. As the synthesis of serotonin and melatonin can be modulated by tryptophan ingestion, the decrease in serotonin and melatonin that normally occurs during aging could be prevented, perhaps some complaints of aging could also be delayed.

In line with this assumption, melatonin has an important role in the aging process due to its ability to reduce oxidative damage due to aging. Besides being a direct scavenger of radicals, melatonin has indirect antioxidative actions as well. All this suggests there may be physiological alterations when melatonin's secretion is reduced on old age. This may be related to the free radical theory of aging.

Oxidative damage is considered a likely cause of age-associated brain dysfunction because the brain is believed to be particularly vulnerable to oxidative stress due to a relatively high rate of oxygen free radical generation without suitable levels of antioxidant defenses compared with other somatic tissues. Exogenous melatonin may prevent the increased production of age related lipid peroxidation products and might have a potential role for retardation of age-related oxidative events.

Recently, the repeated treatment with melatonin or its precursor L-tryptophan, improved the descent in dopamine, serotonin and norepinephrine neurotransmission that normally occurs as a consequence of aging, and it was associated with a restorative effect of the motor and cognitive functions that were impaired as a consequence to aging. Thus, tryptophan intake could be used as a therapy for diseases involving reductions in these neurotransmitters. In this aspect, given the low toxicity of ingested tryptophan and the restrictions that many countries impose against the therapeutic use of melatonin, an adequate dietary supply of tryptophan would be of paramount importance and might aid to improve some age-related degenerative conditions.

Chapter III - Antimicrobial peptides are heavily modified by inclusion of non-native amino acids, peptide cyclization and addition of nonpeptide moieties such as carbohydrates or fatty acids. Peptide acylation can significantly improve the antimicrobial potency of antimicrobial peptides, which primarily target the microbial membrane. This type of antimicrobials, which do not target a specific receptor and whose activity is based on the characteristic lipid composition of microbial membrane has the advantage that it takes microorganism several hundred generations at low concentrations of amphipathic antimicrobial peptide to achieve resistance. In particular, tryptophan and arginine residues have been shown to be almost indispensable due to their amphipathic and cationic character. For instance, arenicin-1, indolicidin, tritrpticin, pleurocidin, melittin, cecropin and so on are having these characteristics. And it also seems to be related to interaction ability in microbial phospholipid membranes. Tryptophan was the first enzyme for which a product formed at one site was demonstrated to be intramolecularly transferred to another site, contributing to substrate channeling. In the history of enzymology and structural biology, tryptophan residue has served a key role because its spatial and functional relationship was deeply investigated. The distinguishing structural characteristic of tryptophan is that it contains an indole functional group. Interestingly, some of antimicrobial peptide showed that tryptophan plays a pivotal role in the membrane-directed antimicrobial activity. The hydrophobic tryptophan residue may support a more efficient interaction with the fungal membrane and bacterial membrane surfaces allowing the peptides to partition into the bilayer interface, in contrast with other hydrophobic residues. Furthermore, chemicals involved indole functional group also possessed membrane interaction ability with antimicrobial effects.

Chapter IV - Amino acids are a major energy source during fish larval stage. Dietary amino acids imbalances have been described when fish larvae are fed rotifers, which may lead to a reduction in the growth rate. Amino acid imbalances may be overcome using amino acid supplementation. There are several ways to assess the effect of amino acid supplementation. The effect of amino acid supplementation can be tested at long term through zootechnique trials or at short term using the tube-feeding technique. Tryptophan is an indispensable amino acid and a precursor of the neurotransmitter serotonin, which affects food intake and aggression in fish. The levels of tryptophan have also been reported to be relevant to the incidence of fish skeletal malformations. Deformities in the vertebral column in reared fish are common and a constraint to the development certain fish cultures. This review studies tryptophan metabolism in a marine fish species, *Diplodus sargus* and evaluates the effects of tryptophan supplementation on fish larval survival, growth and incidence of skeletal deformities.

Commentary - Tryptophan is an essential amino acid and has been used as a dietary supplement for decades because of its alleged health benefits. Apart from its incorporation into body proteins, tryptophan is the precursor for a wide array of metabolites including serotonin, kynurenine, niacin, kynurenic acid and xanthurenic acid, among others.

However, not all metabolites have been identified, and the effects of these substances still remain to be elucidated. Tryptophan is also subject to bacterial degradation in the large intestine, which also gives rise to a great number of metabolites. Disturbances in tryptophan metabolism have been associated with a several diseases, including psychiatric conditions and systemic disorders.

Tryptophan also seems to play a key role in immunoregulatory processes through the kynurenine pathway, which has received increased attention over the past few years. Due to the complexity of the metabolic pathways, health effects of nutritional tryptophan supplementation remain controversial. The question arises whether dietary supplementation of tryptophan is in fact able to compensate for any shortage in the human body and if so, what the optimal conditions are for this.

A more complete understanding of tryptophan metabolic pathways will be necessary to apprehend effects of such tryptophan supplementation. This paper aims to provide a brief overview of findings of nutritional tryptophan supplementation in humans and to discuss future perspectives in tryptophan nutritional research.

In: Tryptophan
Editors: B. L. Whitley et al.
ISBN 978-1-62100-445-5

Chapter I

Molecular Imprinting for Chiral Separation of Tryptophan: Potentials and Prospects

Costas Kiparissides and Olympia Kotrotsiou
Department of Chemical Engineering, Aristotle University of Thessaloniki and Chemical Process Engineering Research Institute, Thessaloniki, Greece

Abstract

Molecular imprinting is an emerging methodology for the creation of selective recognition sites in synthetic polymers. This technique entails the polymerization of functional monomers in the presence of an important molecule (template). Recent studies have shown that the polymers obtained exhibit a surprisingly high degree of stereo- and regiospecific selectivity, making the commercial use of such tailor-made separation materials in the area of chiral separation of bioactive molecules a realistic possibility. This chapter summarizes the present state of the art of molecular imprinting to generate tailor-made chiral stationary phases (CSPs) for the chiral separation of tryptophan and provides an overview of the major factors involved in the manufacturing process that are crucial to the chromatographic performance of the phases.

INTRODUCTION

Enantiomer resolution is a very interesting field in analytical chemistry, especially in pharmaceutical analysis, because frequently enantiomers of the same compound differ strongly in their physiological activity. Usually, in chiral drugs only one enantiomer provides the desired physiological effect. In many cases, the other enantiomer has no effect or is even harmful. Thus, regulators increasingly demand that chiral drugs are administrated in optically extra pure form [1]. On the other hand, due to the fact that life is essentially constructed using L-amino acids as building blocks there is tremendous interest in pharmaceutical industries to produce pure enantiomers [2, 3]. This has intensified efforts of industrial and academic research devoted to develop techniques which are capable to produce pure enantiomers [4-8].

In order to obtain pure enantiomers immense activities have been devoted to study and apply crystallization processes for this purpose. However, the feasibility of selective crystallization depends strongly on the whims of nature, and specifically on the underlying solid-liquid equilibria. Thus, the application of crystallization for the separation or purification of enantiomers requires detailed knowledge of the corresponding phase diagrams describing the melting behaviour of the two enantiomers ('binary' melting point phase diagram) and/or their solubility behaviour in the presence of a suitable solvent ('ternary' solubility phase diagram) [9].

Also, membrane technologies have been applied for the separation of enantiomers. The membranes used for chiral separation can be liquid, including supported liquid membranes, or solid. The liquid membranes [10-12] containing enantiomer recognizing carriers, such as chiral crown ether or cyclodextrins, show highly enantioselective permeability but usually low durability because of losses of the liquid and carriers. This greatly limits their applications to a large extent. Solid membrane is more stable and therefore a durable enantiomer separation process is possible. Solid membranes for enantioselective processes can be categorized into several types: a) chiral polymer membranes, obtained by polymerization of chiral monomers (e.g., chirally derived polysulfone [13, 14], chiral polysaccharides, e.g., chitosan, polyglutamates, etc. [15-17]), b) membranes with immobilized chiral selectors, such as cyclodextrins, chiral ligands or proteins, e.g., bovine serum albumin (BSA), incorporated into nonchiral membranes [18-20] and c) polymer membranes with immobilized chiral catalysts, mostly enzymes, for kinetic resolution of enantiomers, where the two enantiomers of a racemate are transformed into products at different rates [21-23]. The ideal enantioselective membrane should combine good transport rate with high selectivity.

In particular, for the separation of amino acid enantiomers membrane technology using liquid or solid membrane with chiral selector [24-28] has been applied. In most of the cases, the solute flux was very low since the concentration gradient was the only driving force of the process. Attempts to increase the mass transfer were generally limited by the decrease of the purity of the expected compound in the permeate. In other cases, where BSA was chosen as chiral selector for the enantiomeric separation of racemic tryptophan using both solid membrane, with a free chiral selector, and enantiomeric solid membrane, grafted with a chiral selector, the recognition was strongly controlled by many physicochemical parameters (strong pH-dependency). Therefore, it can be concluded that in order to improve the selectivity, recovery and production rate of the separation process all the physicochemical parameters of the feed solution need to be accurately controlled [29].

Another technique that is extensively employed in the analysis of the enantiomeric composition (enantiomeric excess, optical purity) of chiral compounds, is the enantioselective chromatography. Actually, enantioselective separations have been realised in all possible separation techniques, including gas chromatography (GC), column liquid chromatography (CLC), thin-layer chromatography (TLC), as well as electromigration methods, counter current liquid chromatography and liquid-liquid extractions. Liquid and supercritical fluid chromatography (SFC) is also used for the isolation of chiral compounds from racemic mixtures on a preparative scale. Numerous review papers and special monographs describe the technical details as well as the achievements and potential of these important modern separation techniques [30-35].

At present, the analytical methods predominantly used for enantiomer separation are high performance liquid chromatography (HPLC) and capillary electrophoresis (CE). HPLC can be used either indirectly with chiral derivatization reagents (CDR) or directly with chiral stationary phases (CSPs) or chiral mobile phase additives (CMPA). In the last decades, capillary electrophoresis (CE) and capillary electrochromatography (CEC) have been popularly developed and increasingly applied for the enantioseparation of chiral compounds, the main reason being their high efficiency and low solvent and selector consumption [36].

In particular, CE can be considered as complementary to other analytical techniques, such as LC, for enantioseparation. CE techniques can also be classified in two ways, either indirect CE using chiral derivazation agents or direct CE using chiral selectors as additives to the electrolyte. It is a technique that presents a number of advantages: (1) the amounts of samples and separation buffer are much less than those used in HPLC; (2) usually the chiral selectors are dissolved in the background electrolyte (BGE) and thus the expensive chiral columns are not required; (3) higher efficiencies and shorter analysis times are obtained than in HPLC. On the other hand the disadvantages of CE in comparison to HPLC are the lower reproducibility, the poorer sensitivity and the smaller possibility of preparative applications. Moreover transfer of CE is not always straightforward in view of high uncertainty in quantitative analysis [37, 38].

On the other hand, CEC is considered to be a hybrid method between CE and LC combining the characteristically high-peak efficiency of electrically driven separation methods with the high selectivity of chromatographic stationary phases [39]. As CEC is a relatively young analytical separation technique, more researchers will undoubtedly explore this domain to achieve better separation parameters. CEC can be performed in open-tubular capillaries (OT-CEC) and packed capillaries (P-CEC). In the former method, the chiral selector is covalently attached or coated on the inner surface of a capillary; in the latter one, a CSP or an achiral stationary phase in combination with a chiral mobile phase can be used for chiral separations. During the past years, some reviews were reported on this technique [40, 41]. The chiral recognition mechanisms are also based on interactions such as hydrogen bonding, π–π interactions and some forces as those occurring in HPLC [42].

In all these analytical techniques, the crucial factor for chiral separations is the choice of the proper chiral selector, which always needs to be properly carried out according to the structure of the compounds to be analyzed. Hence, the research for and the development of novel chiral selectors are most important to enantioseparation analysis in the present. It is expected that the development and improvement of the separation techniques will solve an increasing number of difficult chiral separation problems and extend the applicational scope of chiral analysis.

Recently, molecularly imprinted polymers (MIPs) have become popular materials for the separation of enantiomers, due to their high selectivity and their chemical and physical stability. More specifically, molecular imprinting is a versatile and facile technique for the preparation of synthetic polymers with predetermined molecular recognition sites for a desired template. MIPs can be prepared by various polymerization methods in the presence of template molecules. The cavities created, after the removal of the template, are complementary in shape and alignment regarding their functional moieties to the template molecules. Thus, this technique enables the formation of tailor-made materials, which can be used as chiral stationary phases in HPLC, TLC, LEC, CE, CEC and SFC. The major advantage of MIP chiral stationary phases over their traditional counterparts is their ability to forecast the order of enantiomer elution. This is due to the fact that the template used in the preparation of the MIPs determines which enantiomer will be most strongly retained by the polymers. Additionally, MIPs are highly cross-linked materials that due to their remarkable chemical and mechanical stability are effective under extreme conditions for lengthy period of time. Moreover, there are no requirements for biological materials in the preparation of MIPs, unlike conventional protein-based technology. Finally, MIPs are inexpensive when compared to most chiral stationary phases and synthetic chiral-selectors [43-48].

The scope of this chapter is to present the recent advances in the development of MIPs, emphasizing on their potential application in the resolution of tryptophan, as well as other amino acid, racemates.

Molecular Imprinting for Enantiomeric Separation

Molecular imprinting is a process where a functional monomer and a cross-linker are co-polymerized (i.e., via a free-radical polymerization mechanism) in the presence of a target molecule (i.e., the so-called imprint molecule) that acts as a molecular template. Initially, the monomers are self-assembled around the template molecule through covalent or non-covalent molecular interactions of the functional groups attached to both the template and the functional monomer. This is followed by the copolymerization of the functional monomer with the bifunctional or trifunctional cross-linker. Subsequent removal of the imprint molecules reveals specific binding sites that are complementary in size and shape to the template molecule [49]. Thus, molecular imprinting is an efficient method for the synthesis of polymers with highly specific recognition sites [50-53].

In general, non-covalent imprinting is easy to achieve and applicable to a wide range of template molecules since many of practically important molecules (e.g., biologically active substances, amino acids, peptides, pharmaceuticals, and environmental contaminants) possess polar groups (e.g., hydroxyl, carboxyl, amino and amide) that can non-covalently interact with the functional groups of the monomer. The advantage of non-covalent imprinting is that the procedure is relatively simple and the synthesis of covalent conjugates is not required prior to polymerization. Furthermore, the template is easily removed from the polymer under very mild conditions since it is only weakly bound to the polymer matrix via non-covalent molecular interactions. Additionally, the guest binding - release, which takes advantage of the non-covalent interactions, is fast. Therefore, non-covalent molecular imprinting has been extensively studied due to its simplicity and versatility [49].

In principle, any kind of non-covalent molecular interactions, including ionic, hydrogen bonding, π-π interactions and hydrophobic interactions, can be effectively employed in molecular imprinting. However, hydrogen bonding is the most appropriate interaction for selective molecular recognition since this type of non-covalent force is highly dependent on both distance and direction between the functional monomer and the template molecule. Thus, monomers that bear the required functional groups (e.g., carboxyl, amino, pyridine, hydroxyl, and amide groups) complementary to the template molecule have been commonly chosen for molecular imprinting [54-55].

In 1990, Andersson et al. prepared for the first time molecular imprints towards enantiomer derivatives of tryptophan utilizing only bonds weaker than covalent and ionic between the print molecule and functional monomers. Methacrylic acid was used as the functional monomer because the acid function of the monomer forms hydrogen bonds with a variety of polar functionalities, such as carboxylic acids, carbamates, heteroatoms and carboxylic esters, of the print molecule. Thus, the interactions, during both the polymerization and the subsequent recognition event, were based solely on hydrogen bonds and other weak forces, such as hydrophobic interactions and dipole-dipole interactions. The free-radical copolymerization of the functional monomer with the cross-linker was carried out in bulk. The cross-linked polymer was subsequently ground and sieved to obtain the final product in particulate form [8-12]. The resulting polymers were analysed for their ability to separate the enantiomers of print molecules and also for their ability to separate the original print molecule from a mixture of compounds similar in structure. In all cases, the stationary phase resolved the enantiomers, while the enantiomeric resolution of a substrate other than the print molecule was always less than that for the print molecule [56].

Few years later [57], some weakly basic monomers, such as N-vinylimidazole and 2-and 4-vinylpyridine (2 VPy and 4 VPy), have been used as functional monomers for molecular imprinting. Ramstrom et al. demonstrated the great potential of using vinylpyridine in combination with methacrylic acid for the preparation of imprints against multifunctional print molecules. MIPs were prepared against Boc-L-tryptophan and evaluated for their ability to resolve the enantiomers of the print molecule in the chromatographic mode. More specifically, the use of 2 VPy as functional monomer was investigated, both as single functional monomer and in combination with MAA and it was proved that the 2 VPy-MIP resolved the enantiomers of Boc-Trp more efficiently than the MAA-MIP (i.e., higher separation factors (α) were obtained). Also, the combination of the two functional monomers (i.e., MAA-2 VPy-MIP) showed even higher enantiomeric separation power. Several types of terpolymers were prepared, where variations in the molar ratio of functional monomer to print molecule were made. It was proved that the polymer prepared with the highest molar ratio of functional monomer to print molecule gave the best separation ($\alpha = 4.35$). This polymer completely resolved the applied amount of racemate (33 nmol, 10 μg) ($f/g = 1.0$). In order to verify the cooperative action of the functional monomers in the terpolymer, reference polymers were made where pyridine and acetic acids were substituted for 2 VPy and MAA, respectively. In both cases, these reference polymers showed much lower enantioselectivity than that recorded for the terpolymer and each copolymer.

In order to gain more insight into the origin of the recognition properties of the imprinted polymers towards tryptophan, a series of experiments have been conducted [58] focusing on the influence of the cross-linker, the functional monomer, and the apparent degree of cross-linking on the recognition site integrity and the subsequent recognition properties of the imprinted polymers. The effect of these parameters on the enantiomeric recognition properties of molecularly imprinted polymers was investigated in the HPLC mode using several types of eluents. The results showed that the imprinted polymer's recognition property is very much influenced by the nature of the polymer network. It was shown that the recognition decreased with a decrease in the apparent degree of cross-linking (molar percentage of cross-linker in the polymerisation mixture). Also, it was shown that the recognition properties of MIPs are greatly influenced by the mobile phase used. In particular, for a polymer prepared in acetonitrile, a good enantiomeric separation was observed when acetonitrile-based mobile phase was used. When the mobile phase was changed to chloroform-based, no enantiomeric recognition was observed although the sample molecule was retarded. This indicated that the specific co-operative binding interactions between the functional groups at the imprinted polymer's recognition sites and the sample molecule were considerably disrupted and only non-specific interactions remained. When the mobile phase was changed back to acetonitrile-based, the recognition was regained. In contrast, for polymers prepared in chloroform, chloroform-based mobile phase gave much better separation than acetonitrile-based mobile phase. When other solvents were tested, significant solvent effects were generally observed. Based on these observations, the recognition properties of the MAA/EGDMA polymers were reinvestigated, and the results showed that by simply using an optimised mobile phase system, significantly improved recognition over previously reported results was observed. So, in optimized conditions for a polymer made against Cbz-L-Trp, 100 μg of Cbz-D,L-Trp was separated with a separation factor (α) of 4.23 and a resolution (R_s) of 3.87, whereas in the previous report, 10 μg of Cbz-D,L-Trp was only separated with α 1.67 and R_s 0.1.

Later on, molecular simulations were carried out to determine the interaction between the functional monomer and the template molecule (i.e., Boc-L-Trp). In this case acrylamide was used as functional monomer. The polymer configurations were first optimized by ab initio prediction. Then the interactions between them were investigated by a molecular dynamics simulation using the software HyperChem 7 [59]. The results showed that when there were one Boc-L-Trp and four acrylamide molecules in a system, the Boc-L-Trp could form 1:3 adducts with acrylamide and that the main interaction between them was hydrogen bonding, which was a valuable result for the optimization of the system in future work.

Also, for the further understanding of the mass transfer mechanisms on MIP stationary phases, the mass transfer kinetics of the Fmoc-L,D-Trp enantiomers on Fmoc-L-Trp imprinted polymer (MIP) and on its reference polymer (NIP), were measured using their elution peak profiles and the breakthrough curves recorded in frontal analysis for the determination of their equilibrium isotherms, at temperatures of 40, 50, 60, and 70°C [60]. At all temperatures, the isotherm data of the Fmoc-Trp enantiomers on the MIP were best accounted for by the Tri-Langmuir isotherm model, while the isotherm data of Fmoc-Trp on the NIP were best accounted for by the Bi-Langmuir isotherm model. The profiles of the elution bands of various amounts of each enantiomer were acquired in the concentration range from 0.1 to 40mM. These experimental profiles were compared with those calculated using the best values estimated for the isotherm parameters and the lumped pore diffusion model (POR), which made possible to calculate the intraparticle diffusion coefficients for each system. The results showed that surface diffusion contributes predominantly to the overall mass transfer kinetics on both the MIP and the NIP, compared to external mass transfer and pore diffusion. The surface diffusion coefficients (*Ds*) of Fmoc-L-Trp on the NIP does not depend on the amount bound (*q*) while the values of *Ds* for the two Fmoc-Trp enantiomers on the MIP increase with increasing *q* at all temperatures. These positive dependencies of *Ds* on *q* for Fmoc-Trp on the MIP were fairly well modeled by indirectly incorporating surface heterogeneity into the surface diffusion coefficient. The results obtained showed that the surface heterogeneity plays an important role in the mass transfer kinetics on imprinted polymers. Thus, in order to improve the chromategraphic performance of imprinted polymers (e.g., to reduce peak tailing), efforts should be directed at decreasing the degree of surface heterogeneity.

These findings, in combination with the fact the preparation of MIPs by bulk polymerization is a process which involves a number of steps (i.e., polymerization, grounding and sieving) that results in a low overall process efficiency, lead to the conclusion that there is a need for the development of alternative methods for the preparation of MIP formats with well-defined morphological characteristics [61-62].

MIP PARTICLES FROM STANDARD POLYMERIZATION TECHNIQUES

The preparation of proper spherical particles can overcome these problems. Spherical MIP micro- and nanoparticles towards tryptophan derivatives have been prepared by using various polymerization methods including suspension, precipitation, miniemulsion, multi-step swelling and grafting (silica or acrylate) of the imprinted polymers (Table 1) [63]. The resulting MIPs, which can be used immediately after preparation and template extraction, are suitable for different applications, including analytical chromatography, solid phase extraction, and other flow-through applications, since columns and cartridges packed with uniform spherical particles exhibit better flow characteristics than those packed with irregular particles.

Suspension Polymerization

In principle, suspension polymerization offers a very attractive alternative to bulk polymers, since by this approach a higher yield of particles with better chromatographic characteristics is produced. In the suspension polymerization process, the organic phase, containing the functional monomer, the cross-linker, the solvent, the template molecule and the chemical initiator, is initially dispersed, with the aid of an agitation system, in the continuous aqueous phase containing the surface-active agent. Subsequently, the temperature is raised to the desired one so that the free-radical copolymerization can be initiated. As the polymerization progresses, the dispersed liquid droplets are gradually transformed from sticky liquid-solid particles into rigid, spherical polymer particles in the size range of 5-500μm [64]. One of the most important issues in suspension polymerization is the control of the droplet/particle size distribution (DSD/PSD). Note that the initial liquid droplet size distribution depends on the type and concentration of the surface-active agent, the quality of agitation and the physical properties (i.e., densities, viscosities and interfacial tension) of the continuous and dispersed phases. Poly(vinyl acetate) partially hydrolyzed to poly(vinyl alcohol) (PVA), is commonly used as stabilizer. The degree of hydrolysis of PVA strongly affects its surface activity at the organic/aqueous interface. In general, the solubility of PVA in the aqueous phase depends on its molecular weight, the sequence length distribution of vinyl alcohol and vinyl acetate monomers in the copolymer chains, the degree of hydrolysis and temperature [65].

Table 1. Comparison of polymerization methods for preparing molecularly imprinted polymer beads [63]

Method	Description	Size Range	Advantages	Disadvantages
Normal suspension polymeriza-tion	Suspension polymeriza-tion in aqueous continuous phase	50-500μm	Normal surfactants may satisfy formation of polymer beads that can be used for column chromatography Highly developed technology with huge literature base Highly reproducible results Large scale possible High-quality beads	Not applicable to imprinting using hydrophilic functional monomers Water is incompatible with most imprinting procedures, only possible for some covalent and metal chelate based processes Phase partition of monomers complicates systems

Method	Description	Size Range	Advantages	Disadvantages
Suspension polymeriza-tion in perfluorocar bon	Suspension polymeriza-tion in perfluorocarb on continuous phase	5-50μm	Applicable to most imprinting systems Particle size adjustable Dispersant does not interfere with imprinting All methods are possible Good quality beads produced	Specialized surfactant needed Expense of liquid fluorocarbons Specialist surfactant polymers required Little literature or optimization to date
Precipitation polymeriza-tion	Cross-linking polymeriza-tion starting from a dilute monomer solution	0.3-10μm	Clean uniform microspheres generally obtained in good yield The microspheres are easy to handle in drug assays	Solvent condition need to be adjusted to minimize consumption of the print molecule
Miniemulsio n polymeriza-tion	Miniemulsion polymeriza-tion	0.1-0.5μm	Generation of small, homogeneous and stable droplets of monomer or polymer precursors, which are then transferred by polymer reactions to the final polymer latexes, without serious exchange kinetics involved	The use of extra stabilizers that probably interfere with the imprinting process
Seed polymeriza-tion	Polymeriza-tion following multi-step swelling of seed particles	-	Uniform polymer beads can be obtained for column chromatography Monodisperse beads excellent packing for HPLC Well established method	Complicated multi-step swelling steps, not generally compatible to non-covalent imprinting systems Need for aqueous emulsions Rules out many imprinting processes Little literature for imprinting
Graft polymeriza-tion	Grafting imprinted polymer layer on supporting beads	-	General compatibility with imprinting, uniform beads can be obtained for column chromatography	Low loading capacity and grafting yield

In previous studies [66], a PVA grade of high molecular weight (i.e., 100,000 MW) and medium degree of hydrolysis (i.e., 86-89%) was selected as stabilizer [67, 68]. It was found that the composition of the organic phase did not significantly affect the final PSD, which was mainly controlled by the concentration of the surface active agent and the degree of agitation. In Figure 1, the measured PSDs of MIP and NIP microparticles, prepared under similar polymerization conditions, are depicted. As can be seen, the two PSDs are almost identical.

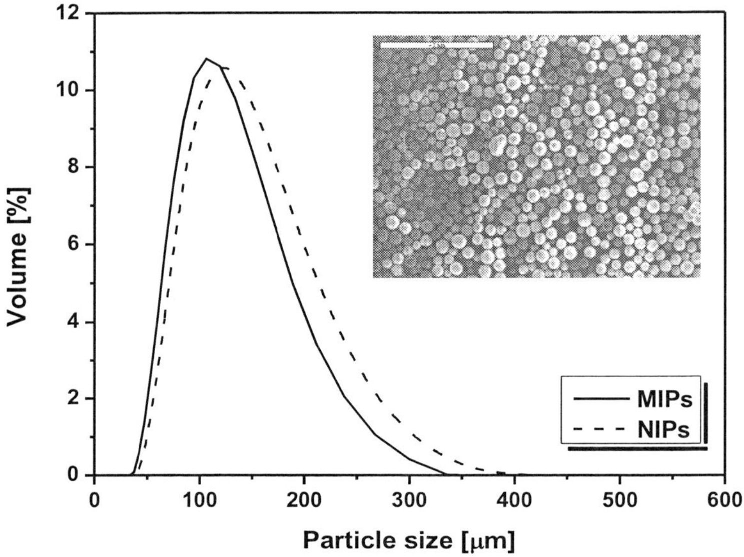

Figure 1. Characteristic particle size distributions of MIP and NIP microparicles.

Also, the effects of process parameters (i.e., porogen concentration, polymerization temperature, types and concentrations of functional monomer and cross-linker) on the particle size distribution and particle morphology were experimentally investigated. In particular, MIP microparticles were prepared using two types of functional monomers (i.e., methacrylic acid (MAA) and methacrylamide (MAm)) and two types of cross-linkers (i.e., ethylene glycol dimethacrylate (EGDMA) and trimethylopropane trimethacrylate (TRIM)). Chloroform dissolved in the dispersed liquid droplets was used as porogen. It was found that the rebinding capacity (selectivity) of the MIP microparticles to the Boc-L-tryptophan molecules was higher than that of the NIPs. Moreover, comparative rebinding experiments of Boc-L-tryptophan and Boc-D-tryptophan enantiomers showed that the binding of Boc-L-tryptophan to the MIP microparticles was higher than that of Boc-D-tryptophan, proving the specificity of synthesized MIPs towards the template molecule. It should be pointed out that the rebinding capacity of the MIP microparticles was further enhanced when the amount of porogen in the dispersed organic phase was increased. That was attributed to the increase of the total pore volume and specific surface area of the microparticles with concomitant the increased accessibility of the specific binding sites [69]. The effects of the cross-linker and functional monomer type on the rebinding capacity of MIPs were also examined (Figure 2).

More specifically, it was found that when the trifunctional cross-linker TRIM was used, the produced MIP microparticles exhibited a higher rebinding capacity and selectivity towards the template molecule.

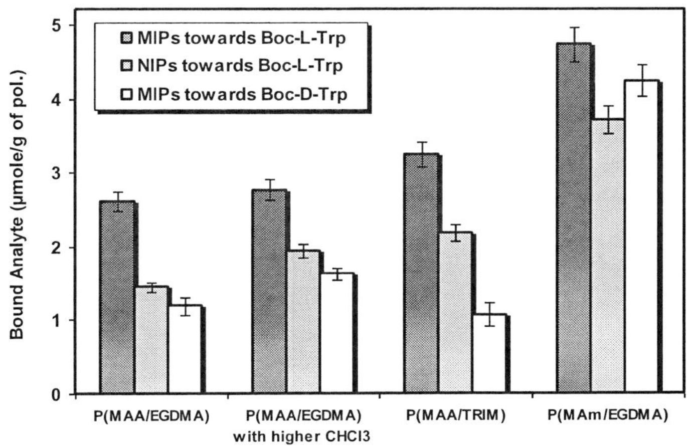

Figure 2. Rebinding capacities of MIPs and NIPs in acetonitrile at 1mM of initial analyte concentration (batchwise guest binding experiments).

This can be explained by the fact that in the presence of TRIM a more rigid polymer matrix, having a higher number of specific binding sites, is obtained as result of the more favorable and efficient arrangement of the template molecules with the functional monomer and the trifunctional cross-linker [70]. With regard to the type of the functional monomer, it was found that the use of MAm, instead of MAA, resulted in a significant increase of the rebinding capacity of MIPs. Although no general agreement on the relative strength of amide and carboxylic acid hydrogen bonds has been reached, it has been suggested that the amide functional groups may be more capable of forming stronger hydrogen bonds even in polar solvents than the carboxyl groups [71]. Alternatively, perfluorocarbons (PFCs), which are largely immiscible with most organic compounds, can also form an appropriate inert phase for suspension polymerization. More specifically, PFC fluids are perfluorinated and saturated aliphatic compounds that include perfluoroalkanes, perfluoroalkylethers and perfluoroalkylamines. Usually, PFCs have higher density, lower boiling point, lower heat of vaporization, and lower polarity than their parent hydrocarbons. Among the most important characteristics are their extremely low miscibility, excellent stability, and inertness toward organic compounds, which make them unmatched by any other fluid used as the universal suspension media. These properties enabled the application of liquid perfluorocarbons in the synthesis of MIPs towards amino acid derivatives [72]. More specifically, a range of polymers imprinted towards Boc-L-Phe (i.e., an amino acid derivative with structure close to that of Boc-L-Trp), was evaluated by HPLC. The results showed excellent resolution (f/g = 0.89, 0.89 and 0.85 at 0.5, 1 and 2 ml/min) and high load capacities with very low back pressures. In addition, the packed columns had low back pressure, and high resolution, even at quite high flow rates (f/g = 0.61 at 5ml/min, back pressure 1300psi). However, in order to create reasonably stable emulsion droplets containing monomers, cross-linkers and print molecules, fluorinated surfactants and a range of surface-active polymers containing fluorinated units must be tested to find a suitable candidate for use as a stabilizer, which in most cases requires synthesis. Also, the high cost of perfuorocarbon remains an unsolved problem.

Precipitation Polymerization

Precipitation polymerization is unique to prepare microspheres of uniform size and shape, which can lead to narrow disperse microspheres free of any added surfactant or stabilizer. This technique starts as homogeneous mixtures of monomer, cross-linker, initiator, and optional solvents. During the polymerization, the growing polymer chains phase-separate from the continuous medium both by enthalpic precipitation, due to unfavorable polymer-solvent interactions and by entropic precipitation, because of cross-linking which prevents the polymer and solvent from freely mixing. Particularly in molecular imprinting precipitation polymerization, as the initial (linear and branched) oligomer radicals (incorporating template-functional monomer complex and cross-linker) grow, they begin to interconnect with each other, aggregate into larger entities and phase-separate from solution while absorbing more template functional monomer complex and cross-linker. In the later stage, new oligomers precipitate onto the preformed nuclei to form enlarged particles, and further polymerization and cross-linking within the monomer-swollen particles yield a highly cross-linked structure, by which specific binding sites are fixed. In good solvents, these polymerizations will often produce turbid macroscopic or microscopic gels, depending largely on the original monomer concentration. In poorer solvents, precipitation polymerization normally produces micrometer-sized particles with narrow size distribution [73]. Thus, in order to synthesize uniform MIP microspheres, the choice of the reaction solvent is critical. More specifically, the solvent should be miscible with monomers and initiator, but be a non-solvent (or non-swelling solvent) for the polymer formed. Especially, at the non-covalent approach the solvent should be non-interfering to the driving force of forming the print molecule-functional monomer complex. Also, the total volume of the solvent for a fixed amount of monomer mixture defines the polymer structure. Precipitation polymerization has also been employed for the synthesis of MIPs using Boc-L-Trp as template molecule [74]. Acetonitrile and chloroform were selected as solvents for the imprinting of Boc-L-Trp, using either MAA or MAm as functional monomers and EGDMA or TRIM as cross-linkers. Figure 3 depicts the effect of the solvent type, the type of functional monomer and cross-linker on the morphology of the Boc-L-Trp imprinted polymers. It is apparent that the use of $CHCl_3$ leads to the formation of nanosponge aggregates, which can be ascribed to the fact that the chlorinated solvents cause severe swelling of the produced polymer [42]. Also, the use of MAA results in the formation of spherical individual particles, whereas the use of MAm as a functional monomer leads to the synthesis of irregular, coagulated particles. This can be explained as follows: when the solubility of the polymer in the reaction solvent is low (e.g., poly(methacrylic acid) (PMAA) in ACN), phase separation occurs at an early stage and leads to the synthesis of nanoporous microspheres with relatively small sizes [75, 76]. Regarding the cross-linker type, it is apparent that the use of a trifunctional cross-linker (e.g., TRIM) leads to the synthesis of smaller particles with slightly rougher surface as compared to those prepared using the bifunctional cross-linking agent EGDMA. This can be attributed to the fact that the use of a cross-linker with multiple vinyl groups, such as TRIM, leads to the formation of more rigid polymeric particles that separate earlier from the reaction medium during the polymerization [77]. For the evaluation of the binding properties of the polymers, batchwise guest-binding experiments were performed in solutions of the template molecule, using as solvent the solvent of the reaction medium (i.e., acetonitrile and chloroform respectively). Competitive analysis was also performed employing Boc-D-Trp in order to examine the selectivity of the Boc-L-Trp imprinted polymeric receptors. The binding capacity/selectivity results of the Boc-L-Trp polymers are presented in Figure 4. As can be observed, the MIP particles adsorb an increased quantity of the template as compared to the NIPs and exhibit a higher affinity for Boc-L-Trp in comparison to its enantiomer Boc-D-Trp, thus proving their selectivity. Concerning the selection of the solvent, it is apparent that when $CHCl_3$ is used, both MIP and NIP particles show good rebinding capacity.

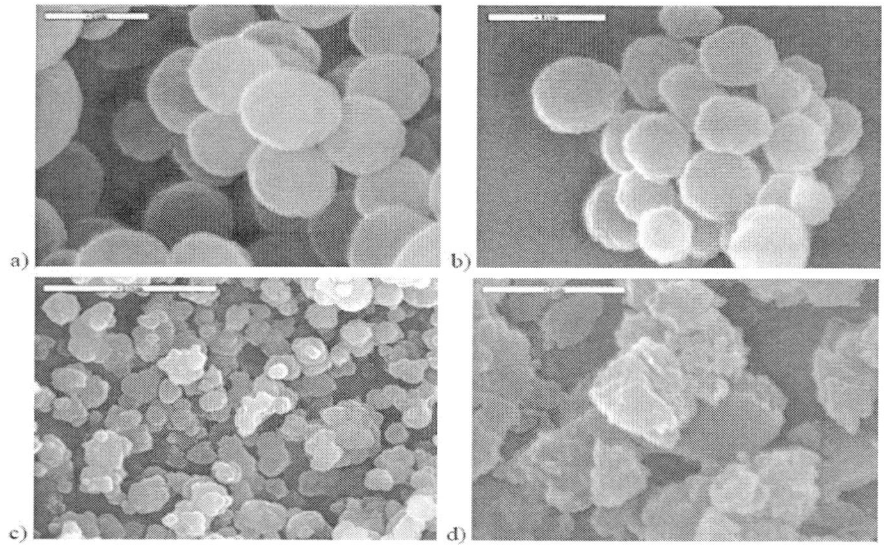

Figure 3. Effect of the functional monomer, cross-linker type and solvent on the MIP particle morphology, (a) P(MAA/EGDMA), (b) P(MAA/TRIM), (c) P(MAm/EGDMA) in ACN and d) P(MAA/EGDMA) in $CHCl_3$.

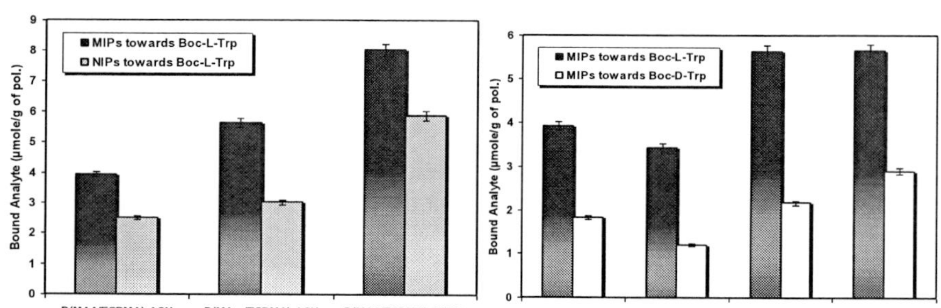

Figure 4. Rebinding capacity of imprinted polymers prepared by precipitation polymerization using a tryptophane-derivative as template molecule.

This can be attributed to the swelling of the polymeric particles promoted by the chlorinated solvent, as mentioned previously, which leads to changes in the three-dimensional configuration of the functional groups of the MIP particles taking part in the recognition and thus results in poor binding specificity [42]. Concerning the effect of the functional monomer, it was proved that the P(MAm/EGDMA) particles show improved binding capacity as compared to the P(MAA/EGDMA) ones. This behavior can be explained by the fact that MAm facilitates stronger non-covalent interactions with the functional groups of the template, which has a direct impact to the rebinding capacity of the MIP particles. Regarding the cross-linker type, it is apparent that in precipitation polymerization the cross-linker has a negligible effect on the rebinding results of MIPs, since the imprinted nanoparticles adsorb approximately the same quantity of the template molecule.

MINIEMULSION POLYMERIZATION

In miniemulsion polymerization the reaction starts from a specially formulated heterophase system (e.g., miniemulsion), where stable nanodroplets of one phase is dispersed in a second, continuous phase. Each of those nanodroplets can be regarded as an individual batch reactor that results into a polymeric nanoparticle at the end of the polymerization reaction with an average diameter in the range of 100-500 nm. Miniemulsions rely on the appropriate combination of saturated high shear treatment, surfactants, and the presence of an osmotic pressure agent insoluble in the continuous phase (e.g., co-surfactant / hydrophobe). Commonly, in miniemulsion polymerization, there are two kinds of co-surfactant that can be used: long-chain alcohols, such as cetyl alcohol, and long-chain alkanes, such as hexadecane. In this polymerization technique the droplet diameter, and consequently the final particle diameter, is adjusted by the type and concentration of surfactant and co-surfactant, the volume fraction of the dispersed phase, the ratio of the oil and aqueous phase, the densities and viscosities of the two phases, the input energy (e.g., volume and duration) and the reaction temperature [78]. In 2002, for the first time the preparation of MIP nanoparticles by miniemulsion polymerisation has been reported [79]. The polymers were prepared using EDMA as cross-linker and MAA as functional monomer (cross-linking degree up to 80 mol %). The template molecule was an N-protected amino acid derivative, i.e., Boc-L-phenylalanine-anilide as well as its enantiomer.

The resulting nanoparticles were 50-300 nm in size in the dry state and exhibited enhanced binding capacity for the template molecule in comparison to nanoparticles imprinted with another enantiomer or to a non-imprinted control. These nanoparticles were used for the preparation of composite membranes. More specifically, a new approach for the development of a selective membrane-based separation has been achieved by coating molecularly imprinted nanoparticles, prepared by miniemulsion polymerization, onto a polyamide membrane-disc. To avoid any loss of particles during the separation process, the coated membrane was covered by a second identical membrane-disc and both membranes were clotted together. This combination of a cost efficient membrane disc as support with a selective nanoparticle layer can provide an efficient way for the enantiomer resolution [80].

MIP Particles from Preformed Seeds

Having to prepare MIPs in beaded form, a possible alternative to their direct synthesis is to use preformed beaded "seeds" as scaffolds for the MIPs to be synthesised. In this connection, different strategies are in principle possible. One of them involves use of porous seeds that can be swollen with the monomer mixture yielding the MIP. Subsequent polymerisation within the seeds yields a composite in beaded form containing the MIP component [81].

The seed can be subsequently removed prior to MIP use or left in the composite. The other strategy involves the generation of a thin MIP layer on the surface of the seed. In this case, the MIP must be formed by grafting polymerization at the seed surface, ideally yielding particles covered with a MIP layer with controllable thickness.

Multistep Swelling and Seed Polymerization

This method involves a combination of stepwise swelling of preformed seed particles followed by polymerization. Aqueous dispersions of organic polymer seeds, usually made out of polystyrene (PS) latexes of ca. 1 μm diameter prepared by emulsifier-free emulsion polymerisation [82] are generally employed for this purpose.

The crucial step is the "activation" [83] of the preformed seed particles by addition of a suitable solvent during the first swelling step. The activation solvent, a water-insoluble, low molecular weight organic compound with plasticising properties, such as dibutyl phthalate, increases the swelling capacity of the seeds up to more than 100 times when introduced before monomers are used to swell the seeds. After the activation, PS seed particles are further swollen by contacting them with an aqueous dispersion containing monomers, template, initiator and optionally a porogenic solvent, in one or more steps, and subsequently polymerised. Particle size, porosity and monodispersity can be controlled by the amount of the activating solvent, porogen, dispersion stabiliser and cross-linker/water ratio; if necessary, a desired particle size can be reached also by repeating several times the swelling and polymerisation steps. In this way, monodisperse polymer particles with sizes larger than those obtainable via direct dispersion/precipitation polymerisation and in the range suitable for direct application as stationary phases can be prepared. To this direction, an attempt has been made to use the aqueous two-step swelling procedure to make monodisperse-imprinted particles towards amino acids.

More specifically, tyrosine-imprinted microspheres have been prepared in an aqueous system by seed swelling procedure, using TRIM, acrylamide (Am) as well as MAA, linear polystyrene and toluene as cross-linker, functional monomers, seed and porogen, respectively [84]. The size distribution proved to be greatly influenced by the ratio of water:TRIM (W/T) and the concentration of dispersant. When W/T was 46:1 (v/v), 4.6% (by mass) of poly(vinyl alcohol) (PVA) was used as dispersant and the molar ratios of tyrosine, MAA, Am and TRIM were 1:8:8:17, the polymer beads had the better size monodispersity, and the average size was 135 μm, while the sizes of the pores on the beads surfaces ranged from 1.25 μm to 9.0 μm. The adsorption behaviour and molecular selectivity of the beads were analysed using liquid chromatography. The results showed that the adsorption behaviour of the beads followed the rule of Langmuir, and the value of saturated absorption was 0.197 mmol g^{-1}. The tyrosine-imprinted polymers exhibited an inherent selectivity for tyrosine, while control polymers had no distinct selectivity for molecule at all.

Graft Polymerization

One other approach is to graft copolymerize the monomer-template assemblies and cross-linker to a reactive support with the desired physico-mechanical properties. In this direction, attempts have been made to produce composite beaded particles, by imprinting in the pore network of silica or preformed beaded poly (trimethylopropane trimethacrylate) particles containing residual double bonds.

In the latter case, prepared porous poly (TRIM) particles via suspension polymerization are used for the production of composite polymers. These composite polymers contain high amounts (up to 64 mol %) of molecularly imprinted cross-linked polymers grafted to the pore system of the spherical microparticulate poly (TRIM) particles (5-10μm in diameter). The resulting composite polymers can be packed into columns directly after washing and then be used in HPLC. The results of previous experimental work have shown that the enantioselectivity for the print molecule Boc-L-Phe was in the same range as traditional bulk phase molecular imprinting, while the performance in HPLC was improved both in terms of efficiency and pressure stability [85].

On the other hand, silica-based supports, which are available in a broad range of bead sizes and pore diameters, are widely used in chromatography. In this case, a class of monomers namely organic silanes are applicable for imprinting since they allow the preparation of substrate-selective siloxane copolymers coated on microparticulate porous silica. These compounds are chosen because a number of organic silanes with a variety of properties are commercially available. Furthermore, since silanes spontaneously form siloxane polymers in aqueous solutions, they may be potentially useful for molecular imprinting of normally water-soluble biomolecules.

Another approach is to graft thin coatings of molecularly imprinted metal complexing polymers to activated silica beads. The "grafting from" approach, which relies on immobilized radical initiators leading to chain growth mainly confined to the support surface, appears particularly promising. In this way, the thickness of the grafted polymer can be better controlled and a higher density of grafted polymer chains can be achieved [86]. By producing thin grafted films the mass transfer problems associated with the majority of MIPs would be largely overcome [87]. These anticipated benefits have been demonstrated by grafting poly(MAA-co-EDMA) imprinted with L-PA on the surface of silica supports modified with azoinitiators. The materials could be prepared in a short time (1-2 hours) and exhibited dramatically superior mass transfer properties when compared to previously described monolith MIPs in liquid chromatography. Thus, 10nm pore size silica containing ultrathin grafted films ($\approx$ 1nm) exhibited the highest chromatographic efficiency with theoretical plate numbers (N) of ca 24000/m for the nonimprinted enantiomer. These results show that the intrinsic efficiency of these phases is acceptable and comparable to conventional columns [88].

The composites were thus suitable as packing material for CEC capillaries [89, 90], since they exhibited enantioselectivies comparable with that showed in LC, high efficiencies, short analysis times (< 3min) and good reproducibility. One problem with using immobilized symmetric azoinitiators is that decomposition leads to one soluble mobile initiating radical in addition to the desired immobilized intiating radical. This results in premature gelation and precludes reuse of the rest unreacted monomers. In order to overcome this problem, initiators producing stable free radicals can be used. In this case, only one of the radicals is capable of initiating polymerization, with the other radical being stable, but capable of terminating the growing polymer chains by recombination. The latter polymerization may exhibit living properties in the sense that a second polymer can be grafted to the first [88]. Recently, it is showed that functional imprinted composite materials could be prepared by grafting of MIPs onto benzyl-N,N-diethyldithiocarbamate iniferter modified porous silica supports [91]. The living properties of the latter system was demonstrated by the consecutive grafting of two polymer layers imprinted with two different enantiomers of the template or one imprinted and one nonimprinted layer in any order [92]. The success of the grafting was reflected in the separation of the two enantiomers obtained using the composite materials as chromatographic stationary phases. Materials containing an initial grafted layer of ca. 16nm prepared in presence of D-PA exhibited pronounced D-selectivity. The living properties of the iniferter were then assessed by grafting of a second layer targeted towards the optical antipode of the template used in the first layer. Thus, grafting of an L-PA imprinted layer on the D-PA imprinted layer resulted in reversal of the enantioselectivity. This shows that different molecular recognition features can be introduced in consecutively grafted layers [88].

MOLECULARLY IMPRINTED MONOLITHS FOR HPLC AND CEC

A major disadvantage regarding the synthesis of MIP particles using standard polymerization techniques, such as suspension, precipitation, miniemulsion polymerization, as well as the techniques involving the utilization of preformed seeds, is the fact that a relatively large quantity of the template molecule is often required for the synthesis of MIPs due to the high dilution factor. Thus, despite the high polymerization yield and the efficiency of the above mentioned techniques, alternative ones have been suggested. More specifically, in situ polymerization techniques have been employed for the preparation of molecularly imprinted monolithic polymer rods. By this technique, the template molecule, the functional monomer, the cross-linker and the initiator are dissolved into a mixture of porogenic solvents. The resulted mixture is degassed and poured into a stainless steel column. After polymerization, the template and the porogenic solvents are simply removed by exhaustive washing with methanol-acetic acid. By this method, the molecularly imprinted polymer is prepared by a simple, one-step, in situ, free-radical polymerization ''molding'' process directly within a chromatographic column without the tedious procedures of grinding, sieving and column packing [93].

Because of the simplicity of the preparation protocol, the high reproducibility of the results, the versatile surface chemistry involved and the rapid mass transport mechanism, monolithic supports have attracted significant interest as stationary phases both in HPLC and CEC. In particular, the great advantage of this method is that provides a highly porous stationary phase. Thus, the mobile phases flow through the stationary phase - adsorbent - with low flow resistance even at high flow rates while convection, which is much more rapid mechanism than diffusion, becomes the dominant mass transport mechanism [94, 95]. Furthermore, the preparation of this type of MIP is more cost-effective because the required amount of the template molecule is much lower.

One of the major MIP monolith matrices that has been extensively investigated is the organic polymer-based system. In this type of matrix, a variety of monomers can be utilized (e.g., MAA, Am, 4 VPy, etc), while the final product is characterized by excellent stability even in different pH environments. To this direction, synthesis of HPLC stationary phases have been realized using stainless-steel chromatographic column tubes (i.e., 150x4mm) by in situ polymerization [96]. Cbz-L-Trp, Fmoc-L-Trp, cinchonine (CN), cinchonidine (CD), quinidine and quinine were chosen as template molecules, MAA and 4 VPy as functional monomers and EDMA as cross-linker. The resultant monolith has led to the complete separation of the enantiomers of amino acid derivatives and diastereomers of cinchona alkaloids, while the backpressure during the process was very low due to the existence of large through-pores. In this way, the separation of the enantiomers and diastereomers was efficient even at elevated flow rates.

Also, the preparation of monolithic MIPs [97] was reported for the chiral separation of nateglinide containing D-phenylalanine unit and its L-enantiomer containing L-phenylalanine. According to this work, an acrylamide-based imprinted monolith, was more enantioselective than an MAA-based imprinted polymer. This was attributed to the formation of higher strength non-covalent interactions (i.e., hydrogen bonds) between the template and monomers even in more polar solvents in case of using acrylamide-based imprinted monolith. Other methods for the synthesis of MIP monolithic stationary phases were also employed [98]. More specifically, MIP monoliths towards (1)-nilvadipine have been prepared with or without the presence of a molecule acting as co-template (i.e., Cbz-L-Trp) and using 4-VP as functional monomer, EDMA as cross-linker and toluene/1-dodecanol as porogen. It was proved that nilvadipine enantiomers could be better separated on the MIP when monoliths were prepared with the addition of the co-template molecule. On the other hand when (1)-nilvadipine was used as the sole template molecule, the same results could not be achieved with the resulted monolithic-imprinted polymer.

Because of the high research interest, further theoretical studies concerning the performance of MIP monolith, i.e., thermodynamic and kinetic properties of the imprinted polymer, have been carried out. These studies have been based on the frontal analysis technique, which offers a dynamic method for accurate and rapid determination of adsorption isotherms from simple breakthrough experiments. Even if this approach has been widely used to determine the adsorption isotherms of packed columns of molecularly imprinted polymers, no attempt at application of this technique to the determination of adsorption isotherms of molecularly imprinted polymer monolithic stationary phases prepared by the in-situ synthesis method had been reported previously.

Thus, the frontal analysis technique has been used to determine the adsorption isotherms of the imprinted polymers monolithic stationary phases. The adsorption behavior of template molecule and its structurally related compounds on the surface of the monolith at different conditions were studied. By fitting experimental data to adsorption isotherm models, the related information of sites distribution and site binding behavior on the surface of molecularly imprinted polymer monoliths were obtained [99-103].

For the evaluation of the results, the thermodynamic properties of the copolymers prepared by in situ polymerization method, using as template molecule Fmoc-L-Trp, were compared with that resulted from bulk polymerization [104]. The affinity energy distributions (AED) were calculated by the expectation maximization method and the adsorption isotherms were determined by the frontal analysis. The comparison of the thermodynamic properties on the above mentioned MIPs proved the coexistence of three types of binding sites at their surface: i) the highest energy sites that adsorb only the imprinted molecule or template, ii) the intermediate energy sites that adsorb both the template and its antipode, although part of them may adsorb only the template and iii) the lowest energy sites that provide nonselective interactions for both the template and its antipode. It was proved that the monolithic MIPs have fewer nonselective sites than the bulk MIPs. These results suggest that from a thermodynamic point of view, monolithic MIPs could provide a better enantiomeric separation.

One other approach for the synthesis and the characterization of the monolithic molecularly imprinted polymers was followed using Cbz-L-phenylalanine as template molecule. Different functional monomers were used for the MIP preparation and the columns were tested with a standard mobile phase of 1% acetic acid in acetonitrile. 4-Vinylpyridine has shown the best separation performance with a resolution factor $R>1$ and thus was chosen for a further improvement of the imprinted stationary phase, which led to a resolution factor $R=2.06$. The columns containing 4-vinylpyridine as functional monomer were further characterised by frontal analysis and peak fitting method. These methods allow the determination of the adsorption isotherm which quantifies the separation potential of the imprinted monoliths [105].

The effects of the flow rate and the concentration of the samples to the dynamic binding capacity of MIP monolith were also examined. To this direction, three kinds of monolithic imprinted columns using cinchonine (CN), Cbz-L-Trp and Fmoc-L-Trp as templates were prepared [106]. For the determination of the dynamic binding capacity, breakthrough curve on the monolithic disk was measured applying the frontal analysis. It was proved that at high flow rates the dynamic binding capacity was almost unaffected even though it was relatively lower at the high flow rates than that at the low flow rates. In addition, the effect of the loading concentration of analytes on the dynamic binding capacity, namely adsorption isotherm, was investigated. A non-linear adsorption isotherm of cinchonine was observed on the molecularly imprinted monolith with cinchonine as template, which might be a main reason to result in the peak tailing of template molecule [107]. Therefore, for keeping the relatively high dynamic binding capacity, a high loading concentration of the sample was not recommended in practice.

In conclusion, molecularly imprinted monolithic stationary phases for the enantiomeric separation of tryptophan have been successfully used in the HPLC and CEC mode. It integrates the high selectivity of MIPs and the merits of monolithic stationary phases. Compared to other types of MIPs, the preparation of imprinted monolithic column is quite simple and cost-effective. Good performance of the stationary phase can be achieved by selection of the appropriate porogenic solvents and careful optimization of the polymerization conditions. Their great advantage in the high-speed analyte separation has been demonstrated. In the future, the methodology of monolithic MIP will be expanded further and can successfully compete with all the other well-established separation technologies [108].

MOLECULARLY IMPRINTED MEMBRANES

Similarly to MIP monoliths, molecularly imprinted membranes (MIMs) are of interest for their highly porous morphology inducing a high permselectivity. Indeed, porous imprinted membranes could overcome the problems associated with the limited accessibility of the recognition sites of the traditional bulk imprinted polymers as well as with the lack of selectivity of usual commercial membranes [81].

The transport properties through a MIP membrane are controlled not only by the sieving (or size exclusion) effect, which is caused by the pore structure, but also by the selective absorption effect, which is resulted by the presence of the specific cavities [109, 110]. As a consequence, the separation performance depends not only on the membrane morphology, (i.e., the barrier pore size and the thickness of the membrane) but also on the efficiency of the molecular recognition. Therefore, according to the porous structure of the polymeric material different selective transport mechanisms across MIP membranes could be distinguished. In particular, for nonporous membranes, the interactions between permeand and membrane dominate transport rate and selectivity; the transport mechanism can be described by the solution/diffusion model. For porous membranes, transport rate and selectivity are mainly influenced by sieving or size exclusion. Nevertheless, interactions of solutes with the membrane (pore) surface may significantly alter the membrane performance. Furthermore, with meso- and macroporous membranes, (selective) adsorption can be used for an alternative separation mechanism. Membrane adsorbers are the most important example [111].

For the preparation of the membranes there are mainly two methods. The first one is the sequential method for the synthesis of composite membranes, which takes advantage of the substrate/base membrane's pore structure (barrier pore size), its layer topology (symmetric vs. asymmetric base membrane) as well as of the location of the MIP. The latter can be situated either on top of the composite membrane ("asymmetric" membrane) or inside the composite membrane ("symmetric" membrane). Thus, according to the thickness of MIP film in the composite membrane, the MIP layer can act either as selective barrier and transport phase, or as an affinity adsorbent.

In the second - simultaneous method for preparation of conventional self supported MIP membranes, two alternatively approaches can be followed, namely, the "traditional" in-situ crosslinking polymerization and the "alternative" polymer solution phase inversion, both in the presence of templates. In both cases, the most important process parameter is the control of membrane thickness in order to have sufficient stability and permeability. Also, the "synchronization" of imprinting and film solidification steps are of critical importance for MIP membrane shape, structure and function [111].

Based on the above mentioned methodologies a hybrid inorganic-organic imprinted membrane have been developed for the enantioseparation of D,L-phenylalanine [112]. The membrane was composed both of an organic phase (i.e., sodium alginate) and a silica phase (i.e., poly(3-aminopropyltriethoxysilane) (APTES)). Imprinted recognition sites for D-phenylalanine were formed in the alginate matrix upon the sol-gel process of APTES occurring in the presence of the template. In this case, it was proved that the selectivity of the imprinted membrane was strongly affected by the APTES content. More specifically, this parameter controlled both the flexibility of the membrane and the stability of the imprinted cavities, while the transportation mechanism of D-phenylalanine across the imprinted membrane was shown to be a facilitated permeation process.

Also, a dry phase inversion process has been employed for the preparation of chiral membranes using Boc-D-Trp or Boc-L-Trp as print molecule [113, 114]. In this case, the recognition cavities were formed by the fixation of a polymer conformation adopted upon interaction with the template molecule. This is achieved by the evaporation of the polymer solvent (dry phase separation) in the presence of the template molecule. More specifically, polystyrene that was modified with a peptide recognition group (i.e., H–(Gly)$_4$–CONH–$CH_2C_6H_4$–resin) was employed as matrix a for amino acid imprinting. Enantioselectivity and high fluxes have been obtained for the imprinted membrane as compared to the non imprinted ones, since the D-isomer of the template molecule was incorporated into the membrane imprinted by Boc-D-Trp in preference to the corresponding L-isomer and vice versa [115].

In the wet phase inversion process, the solidification step involves the immersion of the polymer blend matrix in a non-solvent for precipitation or coagulation. Recently, researchers have devoted a great deal of effort to the optical resolution of amino acids, using membranes prepared by wet phase inversion method. To this direction, L-phenylalanine (L-Phe)- imprinted nylon 6, nylon 6,6 and terephthalic phenylene polyamide (TPPP) membranes have been prepared [116] for the optical resolution of Phe. The selectivity of these membranes as evaluated by batch-wise guest binding experiments was 6.8, 4.2 or 1.7 depending on the polymer matrix. Also, a chiral separation of phenylalanine was performed using an L-Phe-imprinted chitosan (CS)/γ - glycidoxypropyltrimethoxysilane hybrid membrane [117]. However, the permselectivity of L-Phe to D-phenylalanine (D-Phe), as measured after the fixation of the membrane into a diffusion cell, was 0.22, thus indicating that D-Phe could pass more easily through the membrane than the template molecule [81].

Also, in a continuation of previous studies on the development of MIPs for water-soluble templates (D- and L-Phe) [118-123], the preparation of poly[(acrylonitrile)-co-(acrylic acid)] (AA/AN) MIMs by a wet phase inversion method was reported. The in situ implantation of a template was done by a noncovalent interaction for the optical resolution of D- and L-Phe. SEM analysis revealed that the membranes were nanoporous and very thin. Ultrafiltration experiments were performed in order to evaluate the sorption ability, binding and the permeation selectivity of membranes, with regard to the chiral separation of the underivatized Phe aqueous mixtures. It was proved that significantly high selectivities can be achieved in a very short time [124].

CONCLUSIONS

Molecular imprinting is a technique with great potential. One of the most promising applications is their use as CSPs. To this direction, a number of polymer systems have been developed for the imprinting of tryptophan derivatives. Research in the field demonstrates the feasibility of polymerizing MIPs with specific recognition properties for chiral separation of tryptophan derivatives, as well as other amino acid derivatives with structure similar to that of tryptophan (e.g., phenylalanine). One of the advantages of MIP CSPs, in comparison with other CSPs, is that the material is designed directly for a specific molecule and that the elution order of the enantiomers is therefore known.

The goal of any endeavour involving chromatographic separations is to achieve the best possible performance with respect to selectivity, resolution, load capacity and analysis time. Many research efforts on MIPs have therefore focused on improving the chromatographic performance. The use of monodisperse spherical beads, monolithic supports or membranes instead of irregular particles improves the efficiency, and investigations in this direction have already given promising results, thus proving the potential application of MIPs to the chiral separation of tryptophan.

REFERENCES

[1] Blumenstein, J. Chiral drugs: regulatory aspects. In: Crosby J, Introduction, In: Collins A, Sheldrake G, Crosby J. (Eds). Chirality in industry II. Developments in the manufacture and applications of optically active compounds. Chichester: John Wiley and Sons; 1997; 11-18.
[2] Stinson, SC. Chiral pharmaceutical. *Chem. and Eng. News*, 2001 79, 56-61.
[3] Rouhi, AM. Chiral Business. *Chem. and Eng. News*, 2003 81, 45-55.
[4] Lorenz, H; Sheehan, P; Seider-Morgenstern, A. Coupling of stimulated moving bed chromatography and fractional crystallization for efficient enantioseparation, *J. Chromat. A.* 2001 908, 201-214.
[5] Maier, NM; Franco, P; Lindner, W. Separation of enantiomers: needs, challenges, perspectives. *J. Chromat. A*, 2001 906, 3-33.
[6] Francotte, E. Enantioselective chromatography as a powerful alternative for the preparation of drug enantiomers. *J. Chromat. A*, 2001 906, 379-397.
[7] Schulte, M; Strube, J. Preparative enantioseparation by simulated moving bed chromatography. *J. Chromat. A*. 2001 906, 399-416.
[8] Juza, M; Mazzotti, M; Morbidelly, M. Simulated moving bed chromatography and its application to chirotechnology. *Trends in Biotechnology*. 2000 18, 108-118.
[9] Lorenz, H; Caplal, F; Polenske, D; Elsner, MP; Seidel-Morgenstern A. Crystallization based separation of enantiomers. *Journal of the University of Chemical Technology and Metallurgy*. 2007 42, 5-16.
[10] Stella, D; Calzado, JA; Girelli, AM; Canepari, S; Bucci, R; Palet, C; Valiente, M. Liquid membranes for chiral separations. Application of cinchonidine as a chiral carrier. *J. Separation Sci.* 2002 25, 229-238.
[11] Pickering, PJ; Chaudhuri, JB. Equilibrium and kinetic studies of the enantioselective complexation of (D/L)-phenylalanine with copper (II) N-decyl-(L)-hydroxyproline. *J. Chem. Eng. Sci.* 1996 52, 377-386.
[12] Bryjak, M; Kozlowski, J; Wieczorek, P; Kafarski, P. Enantioselective transport of amino acid through supported chiral liquid membranes. *J. Membr. Sci.*, 1993 85, 221-228.
[13] Gumi, T; Minguillon, C; Palet, C. Separation of propranolol enantiomers through membranes based on chiral derivatized polysulfone. *Polymer.* 2005 46, 12306-12312.

[14] Yoshikawa, M; Murakoshi, K; Kogita, T; Hanaoka, K; Guiver, MD; Robertson, GP. Chiral separation membranes from modified polysulfone having myrtenal-derived terpenoid side groups. *European Polym. J.* 2006 42, 2532-2539.

[15] Thoelen, C; De bruyn, M; Theunissen, E; Kondo, Y; Vankelecom, IFJ; Grobet, P; Yoshikawa M; and Jacobs, PA. Membranes based on poly(γ-methyl-L-glutamate): synthesis, characterization and use in chiral separations. *J. Membr. Sci.* 2001 186, 153–163.

[16] Kim, JH; Jegal J; and Lee, KH. Optical resolution of α-amino acids through enantioselective polymeric membranes based on polysaccharides. *J. Membr. Sci.* 2003 213, 273–283.

[17] Swapnali, H. Enantioselective permeation of racemic alcohol through polymeric membrane. *J. Membr. Sci.* 2008 310, 174-183.

[18] Snyder, SE; Carey, JR; Pirkle, WH. Biphasic enantioselective partitioning studies using small-molecule chiral selectors. *Tetrahedron.* 2005 61, 7562-7567.

[19] Xiao, Y; Chung, TS. Functionalization of cellulose dialysis membranes for chiral separation using beta-cyclodextrin immobilization. *J. Membr. Sci.* 2007 290, 78–85.

[20] Matsuoka, Y; Kanda, N; Lee YM; and Higuchi, A. Chiral separation of phenylalanine in ultrafiltration through DNA-immobilized chitosan membranes. *J. Membr. Sci.* 2006 280, 116–123.

[21] Choi, SD; Kim, GJ. Enantioselective Hydrolytic Kinetic Resolution of Epoxides Catalyzed by Chiral Co(III) Salen Complexes Immobilized in the Membrane Reactor. *Catalysis Letters,* 2004 92, 35-40.

[22] Wu, DR; Cramer SM; and Belfort, G. Kinetic resolution of racemic glycidyl butyrate using a multiphase membrane enzyme reactor: Experiments and model verification. *Biotechnol. Bioeng.* 1993 41, 979–990.

[23] Ceynowa, J; Koter, I. Methyl-β-cyclodextrin assisted enantioselective ester hydrolysis catalyzed by lipase immobilized in a polymer membrane. *Sep. Sci. Technol.* 2001 36, 2885-2898.

[24] Dzygiel, P; Wieczorek, P; Jonsson, JA; Milewska, M; Kafarski, P. Separation of amino acid enantiomers using supported liquid membrane extraction with chiral phosphates and phosphonates. *Tetrahedron.* 1999 55, 9923–9932.

[25] Tone, S; Masawaki, T; Hamada, T. The optical resolution of amino acids by ultrafiltration fixed with plasma polymerized 1-menthol. *J. Membr. Sci.* 1995 103, 57–63.

[26] Nakamura, M; Kiyohara, S; Saito, K; Sugita, K; Sugo, T. Chiral separation of dl-tryptophan using porous membranes containing multi-layered bovine serum albumin cross-linked with glutaraldehyde. *J. Chromatogr. A.* 1998 822, 53–58.

[27] Kiyohara, S; Nakamura, M; Saito, K; Sugita, K; Sugo, T. Binding of dl-tryptophan to BSA adosrbed in multilayers by polymer chain grafted onto a porous hollow-fiber membrane in a permeation mode. *J. Membr. Sci.* 1999 1521, 43-149.

[28] Higuchi, A; Hara, M; Horiuchi, T; Nakagawa, T. Optical resolution of amino acids by ultrafiltration membranes containing serum albumin. *J. Membr. Sci.* 93 (1993) 157–164.

[29] Randona, J; Garnier, F; Roccaa, JL; Maısterrena B. Optimization of the enantiomeric separation of tryptophan analogs by membrane processes. *Journal of Membrane Science* 2000 175, 111–117.

[30] Davankov, VA; Kurganov, AA; and Bochkov, AS. Resolution of racemates by high performance liquid chromatography. *Adv. Chromatogr.* 1983 22, 71–116.

[31] Armstrong, DW; and Han, SM. Enantiomeric Separations in Chromatography, *CRC Critical Reviews in Analytical Chemistry*, 1988 19, 175–224.

[32] Pirkle, WH; and Pochapsky, TC. Consideration of chiral recognition relevant to the liquid chromatographic separation of enantiomers. *Chem. Rev.*, 1989 89 347–362.

[33] Allenmark, S. Chromatographic Enantioseparation. 2nd ed Ellis Horwood, New York, 1991.

[34] Francotte, E. Contribution of preparative chromatographic resolution to the investigation of chiral phenomena. *J. Chromatogr. A.* 1994 666, 565–601.

[35] Davankov, VA. Analytical chiral separation methods. *Pure and Applied Chemistry.* 1997 69, 1469–1474.

[36] Wang, Z; Ouyang, J; Baeyens WRG. Recent developments of enantioseparation techniques for adrenergic drugs using liquid chromatography and capillary electrophoresis: A review. *Journal of Chromatography B*, 2008 862, 1–14.

[37] Bressolle, F; Audran, M; Pham, TN; Vallon, JJ. Cyclodextrins and enantiomeric separations of drugs by liquid chromatography and capil-

lary electrophoresis: basic principles and new developments. *J. Chromatogr. B* 1996 687, 303-336.
[38] Ha, PTT; Hoogmartens, J; Van Schepdael, AV; Recent advances in pharmaceutical applications of chiral capillary electrophoresis. *J. Pharm. Biomed. Anal.* 2006 41, 1-11.
[39] Scriba, GKE. Pharmaceutical and Biomedical Applications of Chiral Capillary Electrophoresis and Capillary Electrochromatography: An Update. *Electrophoresis.* 2003 24, 2409-2421.
[40] Gubitz, G; Schmid, MG. Chiral separation by chromatographic and electromigration techniques. A Review. *Biopharm. Drug Dispos.* 2001 22, 291-336.
[41] Gubitz, G; Schmid, MG. Recent progress in chiral separation principles in capillary electrophoresis. *Electrophoresis.* 2000 21, 4112-4135.
[42] Sellergren, B. Molecularly Imprinted Polymers–Man-Made Mimics of Antibodies and their Application in Analytical Chemistry, Elsevier, Amsterdam, 2001.
[43] Alexander, C; Andersson, HS; Andersson, LI; Ansell, RJ; Kirsch, N; Nicholls, IA; O'Mahony, J; Whitcombe, MJ. Molecular imprinting science and technology: a survey of the literature for the years up to and including. *J. Mol. Recognit.* 2006, 19, 106–180.
[44] Sellergren, B. Imprinted chiral stationary phases for HPLC. *J. Chromatogr. A*. 2001 906, 227–252.
[45] Ansell, R. Molecularly imprinted polymers for the enantioseparation of chiral drugs, *J. Adv. Drug Deliv. Rev.* 2005 57, 1809–1835.
[46] Remcho, VT; Tan, ZJ. MIPs as chromatographic stationary phases for molecular recognition. *Anal. Chem.* 1999 71, 248A–255A.
[47] Haupt, K. Molecularly imprinted polymers in analytical chemistry. *Analyst.* 2001 126, 747–756.
[48] Khan, H; Khan, T; Park, JK. Separation of phenylalanine racemates using d-phenylalanine imprinted microbeads as HPLC stationary phase. *Separation and Purification Technology.* 2008 62, 363–369.
[49] Komiyama, M; Takeuchi, T; Mukawa, H; Mukawa, T; Asanuma, H. Molecular Imprinting: From Fundamentals to Applications, Wiley, Weinheim, 2003.
[50] Kempe, M; Mosbach, K. Molecular imprinting used for chiral separations. *J. Chromatogr. A.* 1995 694, 3-13.
[51] Nilsson, K; Sakaguchi, K. Molecular Imprinting of Acetylated Carbohydrate Derivatives into Methacrylic Polymers. *J. Chromatogr. A* 1995 707, 199-203.

[52] Wulff, G. Molecular Imprinting in Cross-Linked Materials with the Aid of Molecular Templates - A Way towards Artificial Antibodies. *Angew. Chem. Int. Ed. Engl.* 1995 34, 1812-1832.

[53] Wulff, G; Schanhaff, S. Racemic resolution of free sugars with macroporous polymers prepared by molecular imprinting. Selectivity dependence on the arrangement of functional groups versus spatial arrangements. *J. Org. Chem.* 1991 56, 395-400.

[54] Rachkov, N; Minoura, T. Towards molecularly imprinted polymers selective to peptides and proteins. The epitope approach. *Biochim. Biophys. Acta* 2001 1544, 255-266.

[55] Rachkov, N., Minoura, T. Recognition of oxytocin and oxytocin-related peptides in aqueous media using a molecularly imprinted polymer synthesized by the epitope approach. *J. Chromatogr. A* 2000 889, 111-118.

[56] Andersson, LI; and Mosbach K. Enantiomeric resolution on molecularly imprinted polymers prepared with only non-covalent and non-ionic interactions. *Journal of Chromatography,* 1990 516, 13-322.

[57] Ramstrom, O; Andersson, LA; Mosbach, K. Recognition sites incorporating both pyridinyl and carboxy functionalities prepared by molecular imprinting. *J. Org. Chem.* 1993 58, 7562-7564.

[58] Yu, C; Mosbach K. Influence of mobile phase composition and cross-linking density on the enantiomeric recognition properties of molecularly imprinted polymers. *Journal of Chromatography A,* 2000 888, 63–72.

[59] Sun, R; Yu, H; Luo, H; Shen, Z. Construction and application of a stoichiometric displacement model for retention in chiral recognition of molecular imprinting. *Journal of Chromatography A.* 2004 1055, 1–9.

[60] Kima, H; Kaczmarsk, K. Mass transfer kinetics on the heterogeneous binding sites of molecularly imprinted polymers. *Chemical Engineering Science* 2005 60, 5425 –5444.

[61] Perez-Moral, N; Mayes, AG; Comparative Study of Imprinted Polymer Particles Prepared by Different Polymerisation Methods. *Analyt. Chim. Acta* 2004 504, 15-21.

[62] Ye, L; Cormack, P; Mosbach, K. Molecularly Imprinted Monodisperse Microspheres for Competitive Radioassay. *Anal. Commun.* 1999 36, 35-38.

[63] Ye, L; Cormack, PAG; Mosbach, K; Molecular imprinting on microgel spheres. *Analytica Chimica Acta* 2001 435, 187–196.

[64] Kiparissides, C; Polymerization Reaction Modeling: A Review of Recent Developments and Future Directions. *Chemical Engin. Science* 1996 51, 1637-1659.

[65] Kotoulas, C; and Kiparissides, C; A generalized population balance model for the prediction of particle size distribution in suspension polymerization reactors. Chemical Engin. *Science* 2006 61, 332-346.

[66] Kotrotsiou, O; Chaitidou, S; and Kiparissides, C. Boc-L-Tryptophan Imprinted Polymeric Microparticles for Bioanalytical Applications. *Materials Science and Engineering C* 2008 29, 2141-2146.

[67] Chatzi, EG; and Kiparissides, C. Dynamic Simulationof Bimodal Drop Size Distributions in Low-Coalescence Batch Dispersion Systems. *Chem. Eng. Sci.* 1992 47, 445-456.

[68] Chatzi, EG; and Kiparissides, C. Drop size distributionin high holdup fraction suspension polymerization reactors: Effect of the degree of hydrolysis of PVA stabilizer. *Chemical Engin. Science* 1994 49, 5039-5052.

[69] Farrington, K; Magner, E; Regan, F. Predicting the performance of molecularly imprinted polymers: Selective extraction of caffeine by molecularly imprinted solid phase extraction. *Anal. Chim. Acta* 2006 566, 60-68.

[70] Kempe, M. Antibody-Mimicking Polymers as Chiral Stationary Phases in HPLC. *Anal. Chemistry,* 1996 68, 1948-1953.

[71] Yu, C; Mosbach, K. Molecular Imprinting Utilizing an Amide Functional Group for Hydrogen Bonding Leading to Highly Efficient Polymers. *J. Org. Chem.* 1997 62, 4057-4064.

[72] Mayes, A; Mosbach, K. Molecularly Imprinted Polymer Beads: Suspension Polymerization using a Liquid Perfluorocarbon as Dispersing Phase. *Anal. Chem.* 1996 68, 3769-3774.

[73] Ye, L; Mosbach, K. Molecularly Imprinted Microspheres as Antibody Binding Mimics. *Reactive and Functional Polymers*. 2001 48, 149-157.

[74] Chaitidou, S; Kotrotsiou, O; Kotti, K; Kammona, O; Bukhari M; and Kiparissides, C. Precipitation Polymerization for the Synthesis of Nanostructured Particles. *Materials Science and Engineering B*, 2008 158, 55-59.

[75] Wang, J; Cormack, PAG; Sherrington, DC; Khoshdel, E. Monodisperse, Molecularly Imprinted Polymer Microspheres Prepared by Precipitation Polymerization for Affinity Separation Applications. *Angew. Chem. Int. Ed.*, 2003 42, 5336-5338.

[76] Ciardelli, G; Cioni, B; Cristallini, C; Barbani, N; Silvestri, D; Giusti, P. Acrylic Polymeric Nanospheres for the Release and Recognition of Molecules of Clinical Interest. *Biosens. Bioelectron.* 2004 20, 1083-1090.
[77] Ye, L; Weiss, R; Mosbach, K. Synthesis and Characterization of Molecularly Imprinted Microspheres. *Macromolecules. 2000* 33, 8239-8245.
[78] Antonietti, M; Landfester, K. Polyreactions in Miniemulsions. *Prog. Pol. Sci.* 2002 27, 689-757.
[79] Vaihinger, D; Landfester, K; Krauter, I; Brunner, H; Tovar GEM. Molecularly Imprinted Polymer Nanospheres as Synthetic Affinity Receptors Obtained by Miniemulsion Polymerisation. *Macromol. Chem. Phys.* 2002 203, 1965–1973.
[80] Lehmann, M; Brunner, H; Tovar G.E.M. Selective separations and hydrodynamic studies: a new approach using molecularly imprinted nanosphere composite membranes. *Desalination* 2002 149, 15-32.
[81] Biffis, A; Dvorakova, G; and Falcimaigne-Cordin, A. Physical Forms of MIPs. *2011, 1-54,* DOI: 10.1007/128_2010_110
[82] Smigol, V.; Svec, F.; Hosoya, K.; Wang, Q.; Fréchet, J.M.J. Monodisperse Polymer Beads as Packing Material for High-Performance Liquid Chromatography. Synthesis and Properties of Monodisperse Polystyrene and Methacrylate Latex Seeds. *Angew Makromol. Chem.* 1992 195, 151-164.
[83] Ugelstad, J; Kaggerud, KH; Hansen, FK; Berger, A. Absorption of low molecular weight compounds in aqueous dispersions of polymer-oligomer particles, A two step swelling process of polymer particles giving an enormous increase in absorption capacity. *Makromol. Chem.* 1979 180, 737- 744.
[84] Zhang, L; Cheng, G; Fu, C. Molecular selectivity of tyrosine-imprinted polymers prepared by seed swelling and suspension polymerization. *Pol. Int.* 2002 51, 687-692.
[85] Glad, M; Reinholdsson P; Mosbach, K. Molecularly imprinted composite polymers based on trimethylopropane trimethacrylate (TRIM) particles for efficient enantiomeric separations. *Reactive Polymers.* 1995 25, 47-54.
[86] Prucker, O; Ruhe, J. Mechanism of radical chain polymerization initiated by azo compounds covalently bound to the surface of spherical particles. *Macromolecules* 1998 31, 602-613.

[87] Sulitzky, C; Ruckert, B; Hall, AJ. Grafting of molecularly imprinted polymer films on silica supports containing surface-bound free radical initiators. *Macromolecules* 2002 35, 79-91.

[88] Sellergren, B. Recognition of Enantiomers Using Molecularly Imprinted Polymers. ed. Piletsky, S; and Turner A. Molecular Imprinting of Polymers. Georgetown, Texas U.S.A., Landes Bioscience, 2006.

[89] Quaglia, M; De Lorenzi, E; Sulitzky, C. Surface initiated molecularly imprinted polymer films: A new approach in chiral capillary electrochromatography. *Analyst* 2001 126, 1495-1498.

[90] Quaglia, M; De Lorenzi, E; Sulitzky, C. Molecularly imprinted polymer films grafted from porous or nonporous silica: Novel affinity stationary phases in capillary electrochromatography. *Electrophoresis*. 2003 24, 952-957.

[91] Rueckert, B; Hall, AJ; Sellergren, B. Molecularly imprinted composite materials via iniferter modified supports. *J. Mat. Chem.* 2002 12, 2275-2280.

[92] Rueckert, B; Hall, AJ. Layer by layer grafting of molecularly imprinted polymers via iniferter modified supports. *Adv. Mat.* 2002 14, 1204.

[93] Matsui, J; Kato, T; Takeuchi, T; Suzuki, M; Yokoyama, K; Tamiya, E; Karube, I. Molecular recognition in continuous polymer rods prepared by a molecular imprinting technique. *Anal. Chem.* 1993, 65, 2223–2224.

[94] Yan, HY; Row, KH. Characteristic and synthetic approach of molecularly imprinted polymer *Int. J. Mol. Sci.* 2006, 7, 155–178.

[95] Liu, ZS; Zheng, C; Yan, C; Gao, RY. Molecularly imprinted polymers as a tool for separation in CEC. *Electrophoresis*. 2007, 28, 127–136.

[96] Huang, XD; Zou, HF; Chen, XM; Luo, QZ; Kong, L. Molecularly imprinted monolithic stationary phases for liquid chromatographic separation of enantiomers and diastereomers. *J. Chromatogr. A* 2003 984, 273–282.

[97] Yin, JF; Yang, GL; Chen, Y. Rapid and efficient chiral separation of nateglinide and its l-enantiomer on monolithic molecularly imprinted polymers *J. Chromatogr. A* 2005 1090, 68–75.

[98] Haginaka, J; Futagami, A. Addition of N-carbobenzyloxy-l-tryptophan as a co-template molecule to molecularly imprinted polymer monoliths for (+)-nilvadipine *J. Chromatogr. A* 2008 1185, 258–262.

[99] Li, H; Liu, YJ; Zhang, ZH; Liao, HP; Nie, LH; Yao, SZ. Separation and purification of chlorogenic acid by molecularly imprinted polymer monolithic stationary phase *J. Chromatogr. A* 2005 1098, 66–74.

[100] Kim, H; Guiochon, G. Thermodynamic Studies on the Solvent Effects In Chromatography on Molecularly Imprinted Polymers. 1. Nature of the Organic Modifier *Anal. Chem.* 2005 77, 1708–1717.
[101] Kim, H; Guiochon, G. Thermodynamic Studies on Solvent Effects in Molecularly Imprinted Polymers. 2. Concentration of the Organic Modifier. *Anal. Chem.* 2005 77, 1718–1726.
[102] Kim, H; Guiochon, G. Thermodynamic Studies of the Solvent Effects in Chromatography on Molecularly Imprinted Polymers. 3. Nature of the Organic Mobile Phase. *Anal. Chem.* 2005 77, 2496–2504.
[103] Kim, H; Guiochon, G. Adsorption on Molecularly Imprinted Polymers of Structural Analogues of a Template. Single-Component Adsorption Isotherm Data. *Anal. Chem.* 2005 77, 6415–6425.
[104] Kim, H; Guiochon, G. Comparison of the Thermodynamic Properties of Particulate and Monolithic Columns of Molecularly Imprinted Copolymers. *Anal. Chem.* 2005 77, 93–102.
[105] Seebach, A., Seidel-Morgenstern, A., Enantioseparation on molecularly imprinted monoliths—Preparation and adsorption isotherms. *Anal. Chim. Acta* 2007 591, 57–62.
[106] Huang, XD; Qin, F; Chen, XM; Liu, YQ; Zou, HF. Short columns with molecularly imprinted monolithic stationary phases for rapid separation of diastereomers and enantiomers. *J. Chromatogr. B* 2004 804, 13–18.
[107] Sellergren, B; Shea, KJ. Origin of peak asymmetry and the effect of temperature on solute retention in enantiomer separations on imprinted chiral stationary phases. *J. Chromatogr. A* 1995 690, 29–39.
[108] Zheng, C; Huang, Y; Liu, Z. Recent developments and applications of molecularly imprinted monolithic column for HPLC and CEC, *J. Sep. Sci.* 2011 34, 1–15.
[109] Piletsky, SA; Panasyuk, TL; Piletskaya, EV; Nicholls, IA; Ulbricht, M. Receptor and transport properties of molecularly imprinted polymer membranes – A review. *J. Membr. Sci.* 1999 157, 263-278.
[110] Ulbricht, M. Membrane separations using molecularly imprinted polymers, *J. Chromatogr. B* 2004 804, 113-125.
[111] Ulbricht, M. Molecularly Imprinted Membranes Molecular Imprinting of Polymers. ed. Piletsky, S; and Turner A. Molecular Imprinting of Polymers. Georgetown, Texas U.S.A., Landes Bioscience, 2006.
[112] Wu, H; Zhao, Y; Nie, M; Jiang, Z. Molecularly imprinted organic–inorganic hybrid membranes for selective separation of phenylalanine isomers and its analogue, *Sep. Purif. Technol.* 2009 68, 97-104.

[113] Yoshikawa, M; Ooi, T; Izumi, JI. Novel membrane materials having EEE derivatives as a chiral recognition site. *Eur. Polym. J.* 2001 37 335-342.
[114] Itou, Y; Nakano, M; and Yoshikawa, M. Optical resolution of racemic amino acid derivatives with molecularly imprinted membranes from tetrapetide consisting of glycinyl residues. *J. Membrane Sci.* 2008 325, 371–375.
[115] Yoshikawa, M; Fujisawa, T; Izumi, J. Molecularly imprinted polymeric membranes having EFF derivatives as a chiral recognition site. *Macromol. Chem. Phys.* 1999 200, 1458–1465.
[116] Takeda, K; Abe, M; and Kobayashi, T. Molecular-imprinted nylon membranes for the permselective binding of phenylalanine as optical-resolution membrane adsorbents. *J. Appl. Polym. Sci.* 2005 97, 620–626.
[117] Jiang, Z; Yu, Y; and Wu, H; Preparation of CS/GPTMS hybrid molecularly imprinted membrane for efficient chiral resolution of phenylalanine isomers. *J. Membr. Sci.* 2006 280, 876–882.
[118] Park, JK; and Seo, JI. Characteristics of phenylalanine imprinted membrane prepared by wet phase inversion method. *Korean J. Chem. Eng.* 2002 19, 940–948.
[119] Park, JK; Kim, SJ; and Lee, JW. Adsorption selectivity of phenylalanine imprinted polymer prepared by the wet phase inversion method. *Korean J. Chem. Eng.* 2003 20, 1066–1072.
[120] Park, JK; Khan, H; and Lee, JW. Preparation of phenylalanine imprinted polymer by the sol–gel transition method. *Enzyme Microb. Technol.* 2004 35, 688–693.
[121] Park, JK; and Kim, SJ. Separation of phenylalanine by ultrafiltration using D-Phe imprinted polyacrylonitrile-poly(acrylic acid)-poly(acryl amide) terpolymer membrane. *Korean J. Chem. Eng.* 2004 21, 994–998.
[122] Park, JK; and Lee, JW. Characteristics of selective adsorption using D-phenylalanine imprinted terpolymer beads. *Korean J. Chem. Eng.* 2005 22, 927–931.
[123] Khan, H; and Park, JK. The preparation of D-phenylalanine imprinted microbeads by a novel method of modified suspension polymerization. *Biotechnol. Bioproc. Eng.* 2006 11, 503–509.
[124] Ul-Haq, N; Khan T; and Park JK. Enantioseparation with D-Phe- and L-Phe-imprinted PAN-based membranes by ultrafiltration. *Journal of Chemical Technology and Biotechnology.* 2008 83, 524-533.

In: Tryptophan
Editors: B. L. Whitley et al.
ISBN 978-1-62100-445-5

Chapter II

Tryptophan Ingestion Improves the Synthesis of Serotonin and Melatonin and May Be Related With Delay of Some Conditions of Aging

S. Esteban, C. Garau, S. Aparicio, M. C. Nicolau and R. V. Rial
Institut Universitari de Ciències de la Salut,
University of Balearic Island, Palma, Spain

Abstract

The essential amino acid tryptophan is the precursor in the biosynthesis of the indoleamines serotonin and melatonin. Brain serotonin is synthesized in two reactions beginning with the hydroxylation of tryptophan to 5-hydroxytryptophan which is then decarboxylated to serotonin.

As the hydroxylation of tryptophan is the rate-limiting step in the synthesis of serotonin, tryptophan hydroxylase (TPH) determines the effective concentration of serotonin *in vivo*. TPH is now known to exist in two isoforms: TPH-1 is mainly expressed in the pineal gland and in gut enterochromaffin cells, and TPH-2 is preferentially expressed in the brain.

The TPH activity increases after tryptophan ingestion which shows that the rate-limiting enzyme is far from being saturated in normal conditions, and confirms that the synthesis of serotonin and melatonin can be modulated by tryptophan ingestion.

The administration of L-tryptophan during the light time increases the brain synthesis and metabolism of serotonin. At night, tryptophan's administration led to a smaller increase in the synthesis of serotonin than by day, although the turnover remain unchanged, implying that, in the dark phase, serotonin is used as a substrate for melatonin synthesis. As a consequence, the amount available of this hormone is dependent first on an adequate dietary supply of tryptophan, and second on the balance between serotonin's use as a neurotransmitter and its availability as precursor for melatonin synthesis which is deeply dependent on the environmental light.

As age advances, the nocturnal production of melatonin decreases. It can be partly explained by a decrease of serum/plasma tryptophan concentration in humans related with aging and associated with an enhanced indoleamine (2,3)-dioxygenase activity, which degrades tryptophan to form kynurenine derivatives. As the synthesis of serotonin and melatonin can be modulated by tryptophan ingestion, the decrease in serotonin and melatonin that normally occurs during aging could be prevented, perhaps some complaints of aging could also be delayed.

In line with this assumption, melatonin has an important role in the aging process due to its ability to reduce oxidative damage due to aging. Besides being a direct scavenger of radicals, melatonin has indirect antioxidative actions as well. All this suggests there may be physiological alterations when melatonin's secretion is reduced on old age. This may be related to the free radical theory of aging.

Oxidative damage is considered a likely cause of age-associated brain dysfunction because the brain is believed to be particularly vulnerable to oxidative stress due to a relatively high rate of oxygen free radical generation without suitable levels of antioxidant defenses compared with other somatic tissues. Exogenous melatonin may prevent the increased production of age related lipid peroxidation products and might have a potential role for retardation of age-related oxidative events.

Recently, the repeated treatment with melatonin or its precursor L-tryptophan, improved the descent in dopamine, serotonin and norepinephrine neurotransmission that normally occurs as a consequence of aging, and it was associated with a restorative effect of the motor and cognitive functions that were impaired as a consequence to aging. Thus, tryptophan intake could be used as a therapy for diseases involving reductions in these neurotransmitters. In this aspect, given the low toxicity of ingested tryptophan and the restrictions that many countries impose against the therapeutic use of melatonin, an adequate dietary supply of tryptophan would be of paramount importance and might aid to improve some age-related degenerative conditions.

Tryptophan, Precursor in the Synthesis of Serotonin and Melatonin

The essential amino acid tryptophan is the precursor in the biosynthesis of the indoleamines serotonin (5-HT, 5-hidroxytryptamine) and melatonin (N-acetyl-5-methoxytryptamine). Brain 5-HT is synthesized in two reactions beginning with the hydroxylation of tryptophan to 5-hydroxytryptophan (5-HTP) which is then decarboxylated to 5-HT. As the hydroxylation of tryptophan is the rate-limiting step in the synthesis of serotonin, tryptophan hydroxylase (TPH) determines the effective concentration of serotonin *in vivo*. TPH is now known to exist in two isoforms: TPH-1 is mainly expressed in the pineal gland and in gut enterochromaffin cells (Walther et al. 2003), and TPH-2 is preferentially expressed in the brain where it plays a fundamental role in 5-HT synthesis (Zhang et al. 2004).

It has been reported that either tryptophan administration or a high plasma ratio between tryptophan and large neutral amino acids can raise brain tryptophan levels, approaching the substrate saturation of tryptophan hydroxylase, and accelerating the synthesis and release of serotonin (Wurtman et al., 1980). Other workers find that circulating melatonin levels show a dose- and time-dependent elevation after the administration of tryptophan and 5-HT-releasing drugs to rats (Huether et al., 1993) and men (Hajak et al., 1991), and that the inhibitor of 5-HT synthesis PCPA inhibits melatonin release (Miguez et al., 1977).

Differences were observed when tryptophan ingestion arise at the beginning of light or dark phases in rats. The tryptophan hydroxylase activity increased in both cases (Esteban et al., 2004) which shows that the rate-limiting enzyme tryptophan hydroxylase is far from being saturated in normal conditions, and confirms that the synthesis of serotonin and melatonin can be modulated by tryptophan ingestion. In addition, the difference between the effects of increased tryptophan intake during light and dark phases suggests that tryptophan hydroxylase activity presents circadian fluctuations which seem to be clock controlled. This is coherent with the results of Sun et al.

(2002) who find that 5-HT levels display a marked circadian rhythm that persists in constant darkness.

The administration of L-tryptophan by day increased the brain synthesis and metabolism of 5-HT while tryptophan's administration at night led to a smaller increase in the synthesis of 5-HT than by day, but the turnover remained unchanged, implying that, in the dark phase, 5-HT is used as a substrate for melatonin synthesis which was in agreement with increased levels of melatonin in plasma (Esteban et al., 2004). In the pineal gland, the subcutaneous administration of L-tryptophan at night has been found to increase the synthesis of 5-HT while leaving 5-HIAA levels unaltered (Reiter et al., 1990).

Serotonin is a neurotransmitter involved in many functions throughout the brain, one of which is its fundamental role in the synthesis of melatonin. As a consequence, the amount available of this hormone is dependent first on an adequate dietary supply of tryptophan, and second on the balance between serotonin's use as a neurotransmitter and its availability as precursor for melatonin synthesis which is deeply dependent on the environmental light. Thus, the balance between serotonin's supply and use is profoundly affected by the suppression of melatonin synthesis by day and its activation at night.

Diet and Sleep Quality

There is a clear association between diet and sleep. According to statistics, one in five people, does not enjoy the sleep that everyone needs when it reaches the end of the day. Although there are many causes that can produce sleep disturbances, it is known that diet exerts a significant influence on the quality of sleep. This is partly because some neurotransmitters and hormones related to sleep-wake cycle depends in part on specific nutrients from food (Anderson, 1981). Nutrients that increase the synthesis of neurotransmitters and promoting hormones stimulants of the central nervous system, such as dopamine, adrenaline and nor adrenaline worse sleep, on the contrary, those who favor the release of regulators related to the feeling of relaxation such as melatonin and serotonin induced sleep. As foods that support the sleep, tryptophan is an essential amino acid, component basic protein, necessary for the formation of serotonin and melatonin which are involved in the sleep cycle (Esteban et al., 2004). As a result, disturbances associated with low levels of serotonin (or melatonin) may be treated by administering their dietary precursor, tryptophan, this treatment may increase the amount of the deficient neurotransmitter at synapses and produce clinical benefit (Grawdon et al., 1977). Such efforts to elevate brain neurotransmitter levels with a naturally occurring precursor represent a new approach in medical therapeutics. In the diet, tryptophanis found in foods as milk, bananas, meat especially turkey, fish especially tuna. It is very important that the carbohydrates-bread, rice, pasta and potatoes stimulate insulin secretion hormone that increases the available tryptophan to make serotonin and melatonin. Moreover, vitaminB6 is also required for the biosynthesis of serotonin and may be of importance in diseases with deficiencies in neurotransmitter function (Hartvig et al., 1995).

In addition, inadequate habits can lead to a sleep disorder. Abundant dining and high-protein and fat, among others, increase hydrochloric acid secretion, causing heartburn, discomfort that does not let to sleep well or sleep is interrupted at midnight. But also, stress or anxiety alters the circadian rhythm cycle time by which the organism adjusts the release of hormones and neurotransmitters that regulate sleep-wake times. During sleep there is greater secretion of some hormones such as growth hormone, prolactin, testosterone or Melatonin, and neurotransmitters, especially serotonin. All these substances are involved in regulating phases of sleep and wakefulness.

Tryptophan and Melatonin on Sleep Regulation

Both the synthesis of serotonin and melatonin depend on the availability of L-tryptophan that must be ingested in sufficient quantities. Once assimilated, tryptophan is converted into serotonin and this, in successive stages and only during the dark phase, can be transformed into melatonin. In agreement with these mechanisms, it has been found that increased tryptophan intake is transformed into rises in the quantity of brain serotonin but also in the pineal and plasma melatonin (Cubero et al., 2006; Sanchez et al., 2008; Esteban et al., 2004; Garau et al., 2006a). In fact, the results of increased tryptophan intake are observed as changes in the activity/rest cycle (Garau et al., 2006b). Several basic and clinical studies have demonstrated the hypnotic effect of the amino acid tryptophan when administered at the beginning of the nocturnal rest period. In 1970, for instance, Wyatt and cols., and later Spieweber (1986) showed that tryptophan led to an increase in sleep in humans. It has also been used successfully for years as a therapeutic agent to combat chronic insomnia. The mechanism by which tryptophan raises the circulating levels of the hormone melatonin has been studied in humans (Hajak et al., 1991) and in experimental animals (Huether et al., 1992). However, it is difficult to know whether the effects of tryptophan are due to its activity of serotonin as a neurotransmitter controlling sleep and wakefulness, whether are due to the hypnotic and chronobiotic actions of melatonin, or whether both factors act at the same time. In any case, it is important to note that the administration of L-tryptophan during the light time increases the brain synthesis and metabolism of serotonin. At night, tryptophan's administration led to a smaller increase in the synthesis of serotonin than by day which is used as a substrate for melatonin synthesis. As a consequence the effects of tryptophan administered at the beginning of the dark period can be understood as a melatonin-mediated effect. In good agreement with this assumption, infants fed with a tryptophan enriched milk during the night period showed improvements in the nocturnal sleep parameteres (sleep efficiency, total hours of sleep, sleep latency or nocturnal awakenings) compared to that of infants fed with a standard commercial milk (Aparicio et al., 2007; Cubero et al., 2006, 2007).

But also not forget that the effects of melatonin may depend on the chronotype, i.e. their diurnal (humans) or nocturnal (rodents) habits of experimental subjects. Thus, the administration of both tryptophan and melatonin at the end of light period induced opposite effects on motor activity of nocturnal and diurnal animals, enhanced activity in nocturnal animals, but showed a hypnotic effect and in diurnal species (Aparicio et al., 2006), since melatonin always facilitates the behaviour typical of the dark period (Huber et al., 1998). It has been reported that a target site for the effects of melatonin is the SCN. If the phase of SCN cells relative to local time is the same in animals of different chronotypes, then differences in the timing of overt rhythms must be controlled downstream from the core of the clock, either in the composition of SCN output signal, or in one or more targets of SCN efferents (e.g., presence of interneurons that invert SCN signals in one chronotype relative to the other) (for a review see Mistlberger, 2005). A possible site for circadian signal inversion is the region of the supraventricular zone (sPVZ) (Smale et al., 2003) which projects to sleep–wake circuits in parallel with the SCN, and may be responsible for modifying SCN timing signals to produce nocturnal and diurnal phenotypes. A new hypothesis is based on the hypothermic effects of melatonin which are well recognized in diurnal and nocturnal animals. The differential effects of melatonin in the behaviour of nocturnal and diurnal species could be correlated with the activity of hypothalamic ventrolateralpreoptic (VLPO) nucleus warm (WSN) and cold (CSN) sensitive neurons as well as with their respective sleep-on and wake-on properties. In nocturnal rats, melatonin causes body cooling and enhances wakefulness; conversely, the diurnal low levels of melatonin occur during the warm phase of their circadian cycle of body temperature when they sleep, which fits with the sleep-on properties of the VLPO WSN neurons (Szymusiak et al., 1998). The population of WSN and CSN observed in cats and rats was mixed and the final results of VLPO warming and cooling were dependent on the properties of the majority of each type. However, a small population of wake promoting, warm-sensitive neurons and sleep-promoting cold-sensitive neurons, i.e., neurons with inverse properties have been observed (Alam et al., 1996). As diurnal humans show increased sleep propensity when the circadian temperature cycle is in its descending phase, the nocturnal melatonin secretion should produce body cooling and sleep, while body warming and wakefulness should appear in the absence of melatonin, i.e., during light time. It would involve a simple change in the relative amount of the main pool of VLPO temperature and sleep controlling neurons (Rial et al., 2008).

Tryptophan and Serotonin on Sleep Regulation

It has been widely demonstrated that the levels of brain serotonin undergo important circadian oscillations in correlation with the sleep/wake and activity/rest cycles. The activity of serotonin neurotransmitter in the regulation of sleep-wake cycle is complex; serotonin intervenes in the control of sleep since its rises produce an increase in the proportion of slow wave sleep and, also, the inhibition of its synthesis by para-chlorophenylanine (PCPA) cause permanent insomnia in cats (Petitjean et al., 1885), although its effects are different in other species (Ursin, 2002; Tejada et al., 2011). Paradoxically speaking, serotonin is also fundamental for the production of wakefulness. While the mentioned results from studies using PCPA allowed to assume a permissive role for serotonin in the production of sleep, it implied that without serotonin in the brain, there is no sleep. Nevertheless, after the administration of PCPA in mammals, sleep eventually reappears while brain serotonin was still very low (Dementet al., 1972, see Ursin, 2002 for a review). Moreover, the 5-HT–containing neurons of the reticular formation, the medial and dorsal raphe which innervate the entire forebrain contribute to maintaining a quiet awake state, and the activity of these neurons show maximal firing during wakefulness but no activity during REM sleep (Miller and O'Callaghan, 2006). There are many 5-HT receptor subtypes as well as numerous behaviours attributed to 5-HT transmission, factors that likely have contributed to the confusion regarding the role of this neurotransmitter in sleep/wake states. Serotonergic activity may be accompanied by waking or sleep depending on the brain area and receptor type involved in the response, on the current behavioural state (Ursin, 2002). Indeed, evidence exists to support a role of 5-HT both in sleep and wakefulness. However, a consensus has recently emerged that supports a stronger role in wakefulness because promotion of 5-HT transmission (eg, reuptake inhibition, precursor loading, etc) results in quiet waking (Jones, 2005; Espana and Scammell, 2004). Hence, firing activity of serotonergic dorsal raphe neurons was related to the level of behavioural arousal, since they discharged regularly at a high rate during waking and at progressively slower rates during slow-wave sleep, and ceased firing during REM (Sakai and Crochet, 2001). It seems to be a general tendency that serotonin measured in the extracellular fluid, which presumably mirrors the actual release in the area where it is measured, is highest during waking, reduced during SWS and more reduced during REM sleep (Portas et al., 2001). Changes in the circadian levels of serotonin are essential for the proper regulation of the sleep/wake rhythm, as reflected in the study conducted by Arnulf et al. (2002) in humans which showed that an alteration in nocturnal sleep is the consequence of the decline in diurnal serotonin levels caused by interrupting the ingestion of tryptophan during the morning. It has been suggested that, during wakefulness, serotonin is responsible for initiating a cascade of post-synaptic genomic processes in hypnogenic neurons located in the preoptic area. In this way, the release of serotonin during wakefulness leads to a homeostatic regulation of slow wave sleep (Jouvet, 1999).

Therefore in humans, taking into account the above mentioned, the administration of L-tryptophan during the light time increases the brain synthesis and metabolism of serotonin promoting wakefulness behaviors and leading to a homeostatic regulation of slow wave sleep. At night, tryptophan's administration led to a smaller increase in the synthesis of serotonin than by day which is used as a substrate for melatonin synthesis. As a consequence, the administration of L-tryptophan during the dark time could be used as a therapeutic agent to combat insomnia.

Serotonin, Melatonin and Circadian Function

A role in the endogenous regulation of circadian clocks was also suggested by Glass et al. (2000). Serotonin (5-HT) is an important regulator of the circadian clock located in the SCN. This clock is synchronized by photic and non-photic signals. Light, the principal synchronizer, is received by the retinal ganglion cells and transmitted directly to the SCN through the retinohypothalamic tract (Moore and Eichler, 1972), and indirectly through the geniculohypothalamic tract (Card and Moore, 1989). Non-photic signals arrive to the SCN by a direct serotonergic pathway from mesencephalic raphe nuclei mainly originated from the median raphe nucleus (MRN) (Azmitia and Segal, 1978; Hay-Schmidt et al., 2003) and secondly from the dorsal raphe nucleus (DRN) (Kawano et al., 1996), also they arrive indirectly by serotonergic projections from the dorsal raphe nucleus to the thalamic intergeniculate leaflets which project to the SCN (Azmitia and Segal, 1978).

A circadian rhythm of brain 5-HT release, with higher levels during light phase was reported (Monnet, 2002, Blier et al., 1989). The tryptophan hydroxylase protein levels detected by immunoautoradiography also showed circadian changes in rats (Malek et al., 2004). Variation of 5-HT responses throughout the day has been also described (Klein and Moore, 1979; Martin, 1991). In addition, abundant evidences indicate that melatonin synchronizes various circadian neural and hypothalamic endocrine processes (Dubocovich, 1995). However, the participation of melatonin in the circadian rhythm of 5-HT release has been questioned from in vitro studies in the hypothalamus of rats (Cardinali et al., 1975). Melatonin reduced the spontaneous 5-HT release in slices of rat hippocampus (Monnet, 2002) and also an inhibitory effect of melatonin on the synthesis and release of brain 5-HT was reported (Garau et al., 2006a). Moreover, it was found that melatonin reduced the synthesis and metabolism of 5-HT down to nocturnal levels, suggesting a physiological role of melatonin in the 5-HT circadian synthesis and release. Also, a marked reduction of 5-HT content in the rat hypothalamus was observed after the administration of pharmacological doses of melatonin during the light period (Lin and Chuang, 2002). Different experimental approaches in vitro (Monnet, 2002; Cardinali et al., 1975) and in vivo (Miguez et al., 1994) showed an inhibitory effect of melatonin on 5-HT release during the dark but not during the light phase. Pinealectomy in rats (Miguez et al., 1995) and in birds (Cassone et al., 1983) also modified the release of 5-HT confirming the modulatory role of melatonin.

The age dependent changes that occur in the rhythmic secretion of melatonin and serotonin as organisms age seem to be related with the impairment of sleep quality in old age (Myers and Badia, 1995, Waldhauser et al., 1998). In fact, the age-related phase advance and decline of pineal melatonin production is believed to be due to the degenerative changes of the neural structures (serotonergic and noradrenergic neuron systems) innervating the pineal gland and the suprachiasmatic nuclei rather than to the degeneration of the pineal tissue itself (Ruzsas et al., 2000), and to a decrease in the amount of serotonin (Esteban et al., 2010a, Esteban et al., 2010b). The administration of L-tryptophan to old animals increased 5-HT synthesis, irrespective of the time of administration. This suggests that, tryptophan hydroxylase was always far from being saturated by its tryptophan substrate. Moreover, the metabolite 5-HIAA were strongly increased after L-tryptophan ingestion, but only during light-time suggesting an increased use of the 5-HT for melatonin synthesis at night (Garau et al., 2006a)

Tryptophan, Melatonin and Aging

Besides the well known function of the pineal hormone Melatonin on circadian rhythm transduction namely a transducer of the light signal, melatonin has another important role in the ageing process, as potential drug to relieve oxidative damage, a likely cause of age-associated brain dysfunction. Brain is believed to be particularly vulnerable to oxidative stress compared with other somatic tissues due to several reasons, including its high utilization of O_2 (the source of many damaging species) and its relatively weak antioxidative defense system. Increasing the peroxidizability of dietary fat has an adverse effect on the function of the central nervous system (CNS) by altering membrane function through changes in the concentration of dietary lipid in membrane phospholipids (Eddy and Harman, 1977). Since neurodegenerative changes and dementia are major and debilitating features of aging, curtailing the actions of free radicals and other reactants in the brain is of great importance in potentially alleviating or forestalling cognitive and mental decline in older individuals. In this way, the free radical theory of aging postulated by Harman in 1956 predicts that the production of intracellular reactive oxygen species (ROS) is the major determinant of life span. The classical free radical theory of aging expects that the healthy life span can be increased by minimizing deleterious free radical reactions while not significantly interfering with those essential to the economy of the cells and tissues. Data available indicates that this can be done by keeping body weight down, at a level compatible with a sense of well-being, while ingesting diets adequate in essential nutrients but designed to minimize random free radical reactions in the body. Such diets would contain minimal amounts of components prone to enhance free radical reactions, such as polyunsaturated lipids (Eddy and Harman, 1977), and increased amounts of substances capable of decreasing free radical reaction damage, as antioxidants.

Melatonin has an important role in the aging process due to its ability to reduce oxidative damage due to aging (Reiter et al., 1998). Melatonin and its metabolites have strong antioxidant properties which act as endogenous buffering agents against oxidative stress(Peyrot and Ducrocq, 2008; Manda et al., 2008; Hardeland et al., 2009). Also, besides being a direct scavenger of radicals, melatonin has indirect antioxidative actions as well (Rodriguez et al., 2004; Tan et al., 2007). The role of melatonin in scavenging destructive free radicals was initially uncovered by Tan et al. (1993) who documented that melatonin directly detoxifies the highly damaging hydroxyl radical (•OH). Since then, numerous publications have appeared which have verified melatonin's ability to neutralize a variety of free radicals and related reactants, confirming the ability of melatonin to protect macromolecules from oxidative damage in all subcellular compartments i.e. proteins, lipids and both nuclear and mitochondrial DNA from free radical damage (Reiter et al., 1997, 2001, Tan et al., 2002 among others). Thus, melatonin achieves this widespread protection by means of its ubiquitous actions as a direct free radical scavenger with the ability to detoxify both reactive oxygen and reactive nitrogen species, indirectly increasing the activity of the antioxidative defense systems (potentiation of the redox enzymes). Furthermore, melatonin stimulates a number of antioxidative enzymes including superoxide dismutase, glutathione peroxidase, glutathione reductase, and catalase. Additionally, melatonin experimentally enhances intracellular glutathione (another important antioxidant) levels by stimulating the rate-limiting enzyme in its synthesis, -glutamylcysteine synthase, and also inhibits the proxidative enzymes nitric oxide synthase and lipoxygenase. Most recently, melatonin has been shown to increase the efficiency of the electron transport chain and, as a consequence, to reduce election leakage and the generation of free radicals. Finally, there is evidence that melatonin stabilizes cellular membranes, thereby probably helping them resist oxidative damage. This is especially important during aging (Reiter, 1995, Reiter et al., 1998), as neuronal death during brain aging results, at least in part, from the disruption of synaptic connectivity caused by oxidative stress.

Accordingly, melatonin appears to have neuroprotective properties when administered in a number of models, both in vitro or in vivo (Antolin et al., 2002, Manev et al., 1996, Matsubara et al., 2003). Removal of melatonin by pinealectomy caused the death of hippocampal pyramidal cells while melatonin supplementation then restored this neuroprotectant action and reversed the cell loss (De Butte and Pappas, 2007). Significantly pineal melatonin declines with aging (q2-6) and it has been suggested that age related neurodegenerative diseases, may be causally related to melatonin deficiency.

These multiple actions make melatonin a potentially useful agent in the treatment of neurological disorders that have oxidative damage as part of their etiological basis. Melatonin is more effective than vitamin E (a potent lipid soluble antioxidant) in neutralizing peroxyl radical and is a powerful inhibitor of *in vivo* lipid peroxidation in the brain, probably does so by scavenging the initiating radicals rather than functioning as a chain-breaking antioxidant in the peroxidative cascade. Besides its direct scavenging actions, it indirectly reduces molecular mangling and malfunction by promoting the removal of toxic and potentially toxic agents by stimulating antioxidative enzymes and inhibiting proxidative enzymes. As an additional contribution to the repertoire by which melatonin indirectly limits oxidative destruction, it may stabilize phospholipid interactions in cellular membranes and limit free radical generation at the mitochondrial level by increasing the efficiency of the electron transport chain.

It is well known that as age advances, the nocturnal production of melatonin decreases in human beings and animal (Pang and Tang, 1983; Reiter, 1995; Touitou, 2001; Zhao et al., 2002 among others), and thus all the strong antioxidative properties decrease also, which in turn suggest important physiological alterations. It can be partly explained by a decrease of serum/plasma tryptophan concentration in humans related with aging and associated with an enhanced indoleamine (2,3)-dioxygenase (IDO) activity, which degrades tryptophan to form kynurenine derivatives (Frick et al., 2004). As the synthesis of 5-HT and melatonin can be modulated by tryptophan ingestion, the decrease in 5-HT and melatonin which normally occurs during ageing could be prevented, perhaps some complaints of aging could also be delayed. In recent years, have been published works suggesting that the exogenouous administration of melatonin could prevent the increased production of age-related lipid peroxidation products and might have a potential role for retardation of age-related oxidative events. Dietary manipulations expected to further lower the rate of production of free radical reaction damage, give results in accord with the free radical theory of aging. In this aspect, given the low toxicity of ingested tryptophan and the restrictions that many countries impose against the therapeutic use of melatonin, an adequate dietary supply of tryptophan would be of paramount importance and might aid to improve some age-related degenerative conditions. Recently, the repeated treatment with melatonin or its precursor L-tryptophan, improved the descent in dopamine, serotonin and norepinephrine neurotransmission that normally occurs as a consequence of aging, and it was accompanied to a restorative effect of the motor and cognitive functions that were impaired as a consequence to aging (Esteban et al. 2010a). In addition, it could be also used as a therapy for other diseases involving reductions in these neurotransmitters. In this aspect, given the low toxicity of ingested tryptophan and the restrictions that many countries impose against the therapeutic use of melatonin, an adequate dietary supply of tryptophan would be of paramount importance and might aid to improve some age-related degenerative conditions.

REFERENCES

Alam MN, McGinty D, Szymusiak R. Preoptic/anterior hypothalamic neurons: thermosensitivity in wakefulness and non rapid eye movement sleep. *BrainRes.* 718:76–82, 1996.

Anderson GH. Diet, neurotransmitters and brain function. *Br. Med.Bull.* 37: 95-100, 1981.

Antolin I, Mayo JC, Sainz RM, de los Angeles del Brio M, Herrera F, Martin V, et al. Protective effect of melatonin in a chronic experimental model of Parkinson's disease. *Brain Res.* 943:163–173, 2002.

Aparicio S, Garau C, Esteban S, Nicolau MC, Rivero M,Rial RV. Chrononutrition: Use of dissociated day/night infant milk formulas to improve the development of the wake-sleep rhythms. *Effects of tryptophan. Nutritional Neuroscience* 10(3-4):137-143, 2007.

Aparicio S, Garau C, Nicolau MC, Rial RV, Esteban S. Opposite effects of tryptophan intake on motor activity in ring doves (diurnal) and rats (nocturnal). *Comp.Biochem.Physiol, Part A* 144:173–179, 2006.

Arnulf I, Quintin P, Alvarez JC, Vigil L, Touitou Y, Lèbre AS, Bellenger A, Varoquaux O, Derenne JP, Allilaire JF, Benkelfat C, Leboyer M. Mid-morning tryptophan depletion delays REM sleep onset in healthy subjects. *Neuropsychopharmacology* 27:843-51, 2002.

Azmitia EC, Segal M. An autoradiographic analysis of the differential ascending projections of the dorsal and median raphe nuclei in, the rat. *J. Comp. Neurol.* 79:641–667, 1978.

Blier P, Galzin AM, Langer SZ. Diurnal variation in the function of serotonin terminals in the rat hypothalamus. *J. Neurochem.* 52: 453–459, 1989.

Card JP, Moore RY. Organization of lateral geniculate-hypothalamic connections in the rat. *J. Comp. Neurol.* 284: 135–147, 1989.

Cardinali DP, Nagle CA, Freire F, Rosner JM. Effects of melatonin on neurotransmitter uptake and release by synaptosome-rich homogenates of the rat hypothalamus. *Neuroendocrinology* 18:72–85, 1975.

Cassone VM, Lane RF, Menaker M. Daily rhythms of serotonin metabolism in the medial hypothalamus of the chicken: effects of pinealectomy and exogenous melatonin. *Brain Res.* 289: 129–134, 1983.

Cubero J, Narciso D, Terrón P, Rial R, Esteban S, Rivero M, Parvez H, Rodríguez AB, Barriga C. Chrononutrition applied to formula milks to consolidate infants' sleep/wake cycle. *NeuroEndocrinol. Letters,* 28:360-366, 2007.

Cubero J, Valero V, Narciso D, Rivero M, Marchena JM, Rodríguez AB, Barriga C. L-tryptophan administered orally at night modifies the melatonin plasma levels, phagocytosis and oxidative metabolism of ringdove (Streptopeliaroseogrisea) heterophils. *Mol. Cell Biochem.* 293: 79-85, 2006.

De Butte M., Pappas B.A.. Pinealectomy causes hippocampal CA1 and CA3 cell loss: Reversal by melatonin supplementation. *Neurobiol. Aging*, 28:306–313, 2007.

Dement WC, Mitler MM, Henriksen SJ. Sleep changes during chronic administration of parachlorophenylalanine. *Rev. Can. Biol. 31*(Suppl):239–46, 1972.

Dubocovich ML. Melatonin receptors: are there multiple subtypes? *Trends Pharmacol. Sci.* 16:50–56, 1995.

Eddy DE, Harman D. Free radical theory of aging: effect of age, sex and dietary precursors on rat-brain docosahexanoic acid. *J. Am.Geriatr. Soc.* 25:220-229, 1977.

Espana RA, Scammell TE. Sleep neurobiology for the clinician. *Sleep,* 27:811–20, 2004.

Esteban S, Garau C, Aparicio S, Moranta D, Barceló P, Fiol MA, Rial RV. Chronic melatonin treatment and its precursor L-tryptophan improve the monoaminergic neurotransmission and related behavior in the aged rat brain. *J. Pineal Res.* 48:170-177, 2010a.

Esteban S, Garau C, Aparicio S, Moranta D, Barceló P, Ramis M, Tresguerres JAF, Rial RV. Improving effects of long-term growth hormone treatment on monoaminergic neurotransmission and related behavioral tests in aged rats. *RejuvenationResearch* 13:707-716, 2010b.

Esteban S, Nicolaus C, Garmundi A, Rial RV, Rodríguez A, Ortega,EIbars CB. Effect of orally administered L-tryptophan on serotonin,melatonin and the innate immune response. *Molecular and Cellular Biochemistry* 267:39-46, 2004.

Frick B, Schroecksnadel K, Neurauter G, Leblhuber F, Fuchs D. Increasing production of homocysteine and neopterin and degradation of tryptophan with older age. *Clin. Biochem.* 37, 684–687,2004.

Garau C, Aparicio S, Rial RV, Nicolau MC, Esteban S. Age-related changes in circadian rhythm of serotonin synthesis in ring doves: Effects of increased tryptophan ingestion. *Experimental Gerontology*41: 40-48, 2006 a.

Garau C, Aparicio S, Rial RV, Nicolau MC, Esteban S. Age-related changes in the activity-rest circadian rhythms and c-fos expression of ring Doves with aging. Effects of tryptophan intake. *Experimental Gerontology*,41:430-438, 2006 b.

Glass JD, DiNardo LA, Ehlen JC. Dorsal raphe nuclear stimulation of SCN serotonin release and circadian phase-resetting. *Brain Res.* 859: 224-32, 2000.

Growdon JH, Cohen EL, Wurtman RJ. Treatment of Brain Disease with Dietary Precursors of Neurotransmitters. *Ann. Intern Med.* 86: 337-339, 1977.

Hajak G, Huether G, Blanke J, Blömer M, Freyer C, Poeggeler B, Reimer A, Rodenbeck A, Schulz-Varszegi M, Rüther E. The influence of intravenous L-tryptophan on plasma melatonin and sleep in men. *Pharmacopsychiatry* 24:17-20, 1991.

Hardeland R, Tan DX, Reiter RJ. Kynuramines, metabolites of melatonin and other indoles: the resurrection of an almost forgotten class of biogenic amines. *J. Pineal. Res.* 47:109–126, 2009.

Harman D. The aging process. *Proc. Natl. Acad. Sci. U S A.* 78(11):7124-7128, 1981.

Hartvig P, Lindner K J, BjurlingP, Langstrom B,Tedroff J.. Pyridoxine effect on synthesis rate of serotonin in the monkey brain measured with positron emission tomography. *J. Neural.Transm.* 102:91-97, 1995.

Hay-Schmidt A, Vrang N, Larsen PJ, Mikkelsen JD. Projections from the raphe nuclei to the suprachiasmatic nucleus of the rat. *J. Chem. Neuroanat.* 25:293–310, 2003.

Huber R, Deboer T, Schwierin B, Tobler I. Effect of melatonin on sleep and brain temperature in the Djungarian hamster and the rat. *Physiol.Behav.* 65:77–82, 1998.

Huether G, Poeggeler B, Adler L, Ruther E. Effects of indirectly acting 5-HT receptor agonists on circulating melatonin levels in rats. *Eur. J. Pharmacol.* 238: 249–255, 1993.

Huether G, Poeggeler B, Reimer A, George A. Effect of tryptophan administration on circulating melatonin levels in chicks and rats: evidence for stimulation of melatonin synthesis and release in the gastrointestinal tract. *Life Sci.* 51:945-953, 1992.

Jones BE. From waking to sleeping: neuronal and chemical substrates. *TrendsPharmacol. Sci.* 26:578–86, 2005.

Jouvet M. Sleep and serotonin: an unfinished story. *Neuropsychopharmacology,* 21: 24S-27S, 1999.

Kawano H, Decker K, Reuss S. Is there a direct retina-raphesuprachiasmatic nucleus pathway in the rat? *Neurosci. Lett.* 212:143–146, 1996.

Klein DC, Moore RY. Pineal N-acetyltransferase and hydroxyindole-O-methyltransferase: control by the retinohypothalamic tract and the suprachiasmatic nucleus. *Brain Res.* 174: 245–262, 1979.

Lin MT, Chuang JI. Melatonin potentiates 5-HT(1A) receptor activation in rat hypothalamus and results in hypothermia. *J. Pineal. Res.* 33:14–19, 2002.

Malek ZS, Pevet P, Raison S. Circadian change in tryptophan hydroxylase protein levels within the rat intergeniculate leaflets and raphe nuclei. *Neuroscience,* 125: 749–758, 2004.

Manda K, Ueno M, Anzai K. Space radiation-induced nhibition of neurogenesis in the hippocampal dentate gyrusnd memory impairment in mice: ameliorative potential of he melatonin metabolite, AFMK. *J. Pineal Res* 45:430–438, 2008.

Manev H, Uz T, Kharlamov A, Joo J. Increased brain damage after stroke or excitotoxic seizures in melatonin-deficient rats. *FASEB J.* 10:1546–1551, 1996

Martin K.F. Rhythms in neurotransmitter turnover: focus on the serotonergic system. *Pharmacol. Ther.* 51: 421–429, 1991.

Matsubara E, Bryant-Thomas T, Quinto JP, Henry TL, Poeggeler B, Herbert D, et al. Melatonin increases survival and inhibits oxidative and amyloid pathology in a transgenic model of Alzheimer's disease. *J.Neurochem.* 85:1101–1108, 2003.

Miguez JM, Martin FJ, Aldegunde M. Effects of pinealectomy and melatonin treatments on serotonin uptake and release from synaptosomes of rat hypothalamic regions. *Neurochem. Res.* 20: 1127–1132, 1995.

Miguez JM, Martin FJ, Aldegunde M. Effects of single doses and daily melatonin treatments on serotonin metabolism in rat brain regions. *J. Pineal Res.* 17: 170–176, 1994.

Miguez JM, Martin FJ, Aldegunde M. Melatonin effects on serotonin synthesis and metabolism in the striatum, nucleus accumbens, and dorsal and median raphe nucleus of rats. *Neurochem. Res.* 22: 87–92, 1977.

Miller DB, O'Callaghan JP. The pharmacology of wakefulness. *Metabolism.* 55:S13–19, 2006.

Mistlberger RE. Circadian regulation of sleep in mammals: role of the suprachiasmatic nucleus. *Brain Res.* 49:429–454, 2005.

Monnet FP. Melatonin modulates [3h]serotonin release in the rat hippocampus: effects of circadian rhythm. *J. Neuroendocrinol.* 14:194–199, 2002.

Moore RY, Eichler VB. Loss of a circadian adrenal corticosterone rhythm following suprachiasmatic lesions in the rat. *Brain Res.* 42 201–206, 1972.

Myers BL, Badia P. Changes in circadian rhythms and sleep quality with aging: mechanisms and interventions. *Neurosci.Biobehav. Rev.* 19: 553-71, 1995.

Pang SF, Tang PL. Decreased serum and pineal concentrations of melatonin and N-acetylserotonin in aged male hamsters. *Horm. Res.* 17: 228–234 1983.

Petitjean F, Buda C, Janin M, Sallanon M, Jouvet M. Insomnia caused by administration of para-chlorophenylalanine: reversibility by peripheral or central injection of 5-hydroxytryptophan and serotonin. *Sleep* 8: 56-67, 1985.

Peyrot F, Ducrocq C. Potential role of tryptophan derivatives in stress responses characterized by the generation of reactive oxygen and nitrogen species. *J. Pineal. Res.* 5:235–246, 2008.

Portas CM, Bjorvatn B, Ursin R. Serotonin and the sleep/wake cycle: special emphasis on microdialysis studies. *Prog.Neurobiol.* 60:13–35, 2000.

Reiter RJ, Guerrero JM, Garcia JJ, Acuna-Castroviejo D. Reactive oxygen intermediates, molecular damage and aging. Relation to melatonin. *Ann. NY Acad. Sci.*854:410–454, 1998.

Reiter RJ, King TS, Steinlechner S, Steger RW, Richardson BA. Tryptophan administration inhibits nocturnal N-acetyltransferase activity and melatonin content in the rat pineal gland. *Neuroendocrinology* 52: 291–296, 1990.

Reiter RJ, Ortiz GG, Monti MG, Carneiro RC. Cellular and molecular actions of melatonin as an antioxidant. *Front Horm. Res.* 23:81–88, 1997.

Reiter RJ, Tan DX, Manchester Lc, Qi W. Biochemical reactivity of melatonin with reactive oxygen and nitrogen species: a review of the evidence. *Cell Biochem.Biophys. 34*:237–256, 2001.

Reiter RJ. Oxidative processes and antioxidative defense mechanisms in the aging brain. *FASEB J.* 9:526–533, 1995.

Reiter RJ. Oxidative processes and antioxidative defense mechanisms in the aging brain. *FASEB J.* 9:526–533, 1995.

Rial RV, Akaârir M, Gamundi A, Garau C, Aparicio S,Tejada S, Gené L, Nicolau MC, Esteban S. Wake and sleep hypothalamic regulation in diurnal and nocturnal chronotypes. *Journal ofPineal Research* 45:225-226, 2008.

Rodriguez C, Mayo JC, Sainz RM et al. Regulation of antioxidant enzymes: a significant role for melatonin. *J. Pineal. Res.* 36:1–9, 2004.

Rúzsás C, Mess B. Melatonin and aging. A brief survey. *NeuroEndocrinol.Lett.* 21: 17-23, 2000.

Sakai K, Crochet S. Role of dorsal raphe neurons in paradoxical sleep generation in the cat: no evidence for a serotonergic mechanism. *Eur. J.Neurosci.* 13:103–12, 2001.

Sanchez S, Paredes SD, Sanchez CL, Barriga C, Reiter RJ, Rodriguez AB. Tryptophan administration in rats enhances phagocytic function and reduces oxidative metabolism. *NeuroEndocrinol.Lett.* 29: 1026-1032, 2008.

Smale L, Lee T, Nunez AA. Mammalian diurnality: some facts and gaps. *J. Biol. Rhythms,*18:356–366, 2003.

Spinweber CL. L-tryptophan administered to chronic sleep-onset insomniacs: late-appearing reduction of sleep latency. *Psychopharmacology* (Berl) 90:151-155, 1986.

Sun X, Deng J, Liu T, Borjigin J. Circadian 5-HT production regulated by adrenergic signaling. *Neurobiology,* 99: 4686–4691, 2002.

Szymusiak R, Alam N, Steininger TL, McGinty D. Sleep–waking discharge patterns of ventrolateralpreoptic/anterior hypothalamic neurons in cats. *Brain Res.* 803:178–188, 1998.

Tan DX, Chen LD, Poeggeler B et al. Melatonin: a potent, endogenous hydroxyl radical scavenger. *Endocr. J.* 1993; 1:57–60.

Tan D-X, Manchester LC, Terron MP, Flores LJ, Reiter RJ. One molecule, many derivatives: a never-ending interaction of melatonin with reactive oxygen and nitrogen species? *J. Pineal. Res.* 42:28–42, 2007.

Tan DX, Reiter RJ, Manchester LC, YanMT, El-Sawi M, Sainz RM, et al. Chemical and physical properties and potential mechanism: melatonin as a broad spectrum antioxidant and free radical scavenger. *Curr. Top. Med. Chem.* 2:181–198, 2002.

Tejada S, Rial RV, Gamundí A, Esteban S. Effects of serotonergic drugs on locomotor activity and vigilance states in ring doves. *Behavioral Brain Research*, 216:238-246, 2011.

Touitou Y. Human aging and melatonin. Clinical relevance. *Exp.Gerontol.* 36:1083–1100, 2001.

Ursin R. Serotonin and sleep. *Sleep Med. Rev.* 6: 55-69, 2002.

Waldhauser F, Kovács J, Reiter E. Age-related changes in melatonin levels in humans and its potential consequences for sleep disorders. *Exp.Gerontol.* 33: 759-72, 1998.

Walther DJ, Peter JU, Bashammakh S, Hörtnagl H, Voits M, Fink H, Bader M. Synthesis of serotonin by a second tryptophan hydroxylase isoform. *Science* 299:76, 2003.

Wurtman RJ, Hefti F, Melamed E. Precursor control of neurotransmitter synthesis. *Pharmacol. Rev.* 32: 315–335, 1980.

Wyatt RJ, Engelman K, Kupfer DJ, Fram DH, Sjoerdsma A, Snyder F. Effects of L-tryptophan (a natural sedative) on human sleep. *Lancet* 24:842-846, 1970.

Zhang X, Beaulieu J-M, Sotnikova TD, Gainetdinov RR, Caron MG. Tryptophan hydroxylase-2 controls brain serotonin synthesis. *Science* 305:217, 2004.

Zhao ZY, Xie Y, Fu YR, Bogdan A, Touitou Y. Aging and the circadian rhythm of melatonin: a cross-sectional study of Chinese subjects 30–110 yr of age. ChronobiolInt 19:1171–1182, 2002.

In: Tryptophan
Editors: B. L. Whitley et al.
ISBN 978-1-62100-445-5

Chapter III

Role of Tryptophan Residues in Antimicrobial Activity and Membrane Interactions

In-sok Hwang, Jaeyong Cho, Juneyoung Lee and Dong Gun Lee*

School of Life Sciences and Biotechnology, College of Natural Sciences, Kyungpook National University, Daegu, Korea

Abstract

Antimicrobial peptides are heavily modified by inclusion of non-native amino acids, peptide cyclization and addition of nonpeptide moieties such as carbohydrates or fatty acids. Peptide acylation can significantly improve the antimicrobial potency of antimicrobial peptides, which primarily target the microbial membrane. This type of antimicrobials, which do not target a specific receptor and whose activity is based on the characteristic lipid composition of microbial membrane has the advantage that it takes microorganism several hundred generations at low concentrations of amphipathic antimicrobial peptide to achieve resistance.

*School of Life Sciences and Biotechnology, College of Natural Sciences, Kyungpook National University, Daehak-ro 80, Buk-gu, Daegu 702-701, Republic of Korea, **Tel: 82-53-950-5373; Fax: 82-53-955-5522, E-mail:** dglee222@knu.ac.kr.

In particular, tryptophan and arginine residues have been shown to be almost indispensable due to their amphipathic and cationic character. For instance, arenicin-1, indolicidin, tritrpticin, pleurocidin, melittin, cecropin and so on are having these characteristics. And it also seems to be related to interaction ability in microbial phospholipid membranes. Tryptophan was the first enzyme for which a product formed at one site was demonstrated to be intramolecularly transferred to another site, contributing to substrate channeling. In the history of enzymology and structural biology, tryptophan residue has served a key role because its spatial and functional relationship was deeply investigated. The distinguishing structural characteristic of tryptophan is that it contains an indole functional group. Interestingly, some of antimicrobial peptide showed that tryptophan plays a pivotal role in the membrane-directed antimicrobial activity. The hydrophobic tryptophan residue may support a more efficient interaction with the fungal membrane and bacterial membrane surfaces allowing the peptides to partition into the bilayer interface, in contrast with other hydrophobic residues. Furthermore, chemicals involved indole functional group also possessed membrane interaction ability with antimicrobial effects.

Introduction

During evolution and natural selection, all organisms endowed a vast of tools for survival. One of the systems present as a first line of defence in the innate immune system is antimicrobial peptides. These peptides were initially isolated in the 1980s, from frogs and insects [1]. It has been found to play a vital role in their survival in bacteria infested marsh. Thereafter, a number of additional antimicrobial peptides have been discovered everywhere in nature and there are presently over 800 that have been identified. Most of these peptides possess antimicrobial functions, including antiviral [2], antifungal [3], antitumor [4], and inmmunomodulatory activities [5]. Antimicrobial peptides have been proposed to achieve their bactericidal effect in a number of different fashions. Several of these cases, an initial interaction with the outer or inner membranes of bacteria is essential. This interaction between the peptides and their membranes can occur by a number of different mechanisms. The barrel-stave, toroidal pore, and carpet models are representative [6]. In the barrel-stave model, the peptides span the membrane and form a pore. The toroidal pore model creates pores that contain peptides as well as lipid molecules that are curved inwards towards the pore in a continuous fashion from the surface of the membrane. Lastly, the carpet model is formed a peptide carpet by the peptides lining up parallel to the membrane surface, and then the peptides caused pore formation by a detergent-like action. Not all antimicrobial peptides are thought to affect their major action on membranes. There are many of peptides which described that act on intracellular targets, inhibiting protein synthesis, interact with DNA or RNA, and inhibit some sort of enzymatic activity [7]. One of these breadth activity peptides encompasses an unusual composition of regular amino acids, such Trp residues. Trp residues represent a recurring theme among many antimicrobial peptides and almost play key roles for activity. They also play an important role in membrane spanning proteins, as Trp has a strong preference for the interfacial regions of lipid bilayers [8-10]. In addition, Trp residues are involved in protein folding, forming both native and nonnative hydrophobic contacts even in denatured proteins to ensure their proper folding [11]. Tryptophan is an amino acid required by all forms of life for protein synthesis and other important metabolic functions, but animals do not possess the enzymatic machinery to synthesize it from simpler molecules. At the level of primary producers, tryptophan is synthesized from molecules such as phosphoenolpyruvate in bacteria, fungi and plants, and these organisms fuel the tryptophan flux through the food chain. Because animals are incapable of synthesizing tryptophan, they must ingest it in the form of proteins, which are then hydrolysed into the constituent amino acids in the digestive system.

This article will focus on an importance of Trp residues in antimicrobial effects and examine studies addressing important finding for indolicidin, arenicin-1 and indole-3-carbinol which is a compound based on tryptophan.

ROLE OF TRYPTOPHAN RESIDUES

Trp residues are found in many antimicrobial peptides. Their roles have been investigated extensively but it has yet to determine clearly what specific properties they bring to antimicrobial peptides. In most cases, Trp is considered hydrophobic due to its uncharged sidechain. Another important factor is the extensive π-electron system of the aromatic indole sidechain, which gives rise to a significant quadrupole moment (Figure 1). The quadrupole moment thought to be two dipole moments extending vertically out of either side of the ring plane. Because the Trp sidechains do not have a dipole moment and no other obvious charges, it is often considered as essentially hydrophobic. The π-electron system of the aromatic indole sidechain results in negatively charged clouds that can participate in cation- π interactions that occur in proteins between the negatively charged electron cloud of any aromatic residue and various cationic species, such as ions or positively charged amino acids (Figure 1) [12]. Cation- π interactions play an important role in substrate binding, catalysis, as well as ion channel activity [13].

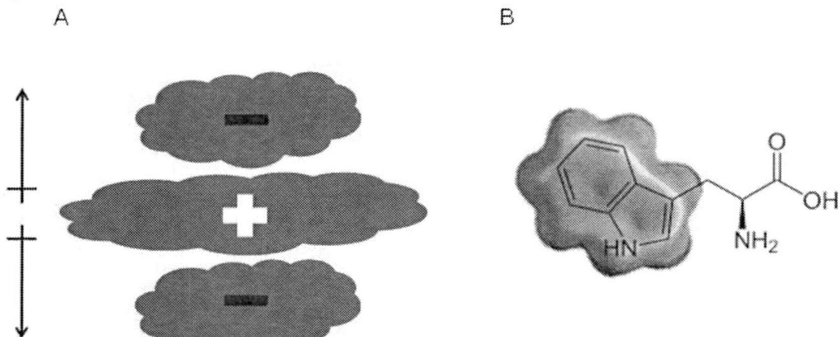

Figure 1. (A) Schematic of the dumbbell shaped π–electron clouds above and below the aromatic ring structure. The arrows on the left indicate dipole moments. (B) Electrostatic surface of the Trip indole group and the line-diagram of the Trip [6].

The fact that Trp residues have a preference for the interfacial region of lipid bilayers is already well known. This phenomenon is not only found in membrane proteins but also observed in many Trp containing antimicrobial peptides. Trp residue can associate with the positively charged choline head groups of the lipid bilayer and it form hydrogen bonds with both water and components of the lipid bilayers when situated in the interfacial region [14, 15]. These hydrogen bonds are not consistently required when the peptide inserts further into the hydrocarbon core.

INDOLICIDIN

Antimicrobial cationic peptides are ubiquitous in nature and are thought to be an important component in innate host defenses against infectious agents [16, 17]. Antimicrobial cationic peptides have been isolated from a wide range of animal, plant, and bacterial species and indolicidin (ILPWKWPWWPWRR), from cytoplasmic granules of bovine neutrophils, has a unique composition consisting of 39% tryptophan and 23% proline.

Indolicidin is the smallest of the naturally known occurring linear antimicrobial peptides, containing the highest percentage of tryptophan of any known protein, and consists of only six different amino acids. Due to the distribution of proline and tryptophan residues throughout the indolicidin sequence, it may assume a structure distinct from the well described a-helical and b-structured peptides. Indolicidin has been shown to be a fairly potent antimicrobial peptide with activity against a variety of microorganisms, fungi, and protozoa. The basis for the generation of bioactive peptides is the understanding of their effective mechanism. While the mechanisms of cationic antimicrobial peptides have not been clearly elucidated, two main theories have been suggested to explain the nature of a possible peptide membrane structure. One is the pore formation in the lipid bilayer by detergent effect due to a common amphipathicstruc ture, and the other is the formation of ion channels by aggregation of several peptide molecules in the lipid bilayer plane. In previously study, in order to elucidate the fungicidal mechanism of the synthetic peptide, indolicidin, we looked into its fungicidal effect on various pathogenic fungal strains. And we discussed the mechanism(s) of indolicidin, with respect to changes in plasma membrane dynamics, and their effectiveness on the salt ion in *Trichosporon beigelii* [18]. The in vitro fungicidal activity of indolicidin was measured against yeast by the MTT assay and expressed as MIC (minimal inhibitory concentration), and the results showed that indolicidin has a remarkable fungicidal activity. In addition, all the fungal strains tested are highly susceptible to indolicidin with MIC values in the 2.5–10 μM range of concentrations (Table 1). Melittin was used to compare with indolicidin. To examine the target site of indolicidin in T. begelii and Aspergillus flavus, they were labeled with FITC and visualized under the confocal microscopy. FITC did not give any affect on the fungicidal activity of indolicidin. The FITC-labeled indolicidin penetrated the cell membrane and accumulated in the plasma membrane of the cell immediately after addition to the cells, suggesting that the major target site of indolicidin is the membrane of the fungal cells. To determine whether the fungicidal effect of indolicidin results from the disintegration of the cell membrane or from an effect on the cell physiology, cells were incubated with the DNA intercalating dye propidium iodide (PI). PI staining of the peptide treated cells was expected to give further information about the cytotoxic mechanism of indolicidin. That is, in case of the disruption of fungal cell membrane, the peptide would be permeabilized and therefore would allow the free diffusion of small dyes such as PI into cytoplasm. Detection of internal PI was analyzed on single cells by FACS analysis.

The result indicated that while untreated normal cells showed no PI fluorescence activity signal, the fluorescence histogram of cells treated with indolicidin and melittin at a concentration of 20 μM showed a total shift of the peak to the right (Figure 2). These cells are regarded as apoptotic or may have spontaneously died during culture, and therefore the fluorescence intensity of these cells serves as a marker for the relative amount of internalized PI in the case of a permeabilized cell membrane like that of disintegrating cells. On the other hand, cells were incubated with indolicidin, at 28 °C for 4 h in the absence or presence of NaN_3 as a respiratory inhibitor.

Table 1. Fungicial activity of indolicidin [18]

Peptides	MIC (μM)		
	S. cerevisae	*T. beigelii*	*C. albicans*
Indolicidin	5–10	2.5–5	5–10
Melittin	1.25	1.25	1.25

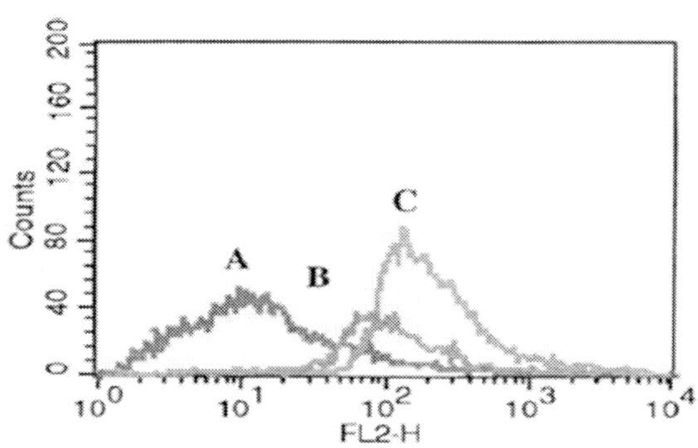

Figure 2. FACScan analysis of propidium iodide staining in *T. beigelii*. *T.beigelii* were mixed with peptides at a concentration of 20 μM. Permeabilization of the cell membrane was detected by incubation of peptide treated cells in propidium iodide. Histograms show the fluorescence intensity of internalized PI after peptide treatment of T. beigelii. (A) Control staining without any peptide treatment; (B) PI staining of cells treated with indolicidin; (C) staining of cells treated with melittin [18].

As shown in Figure 3, the interaction of peptides with T. beigelii cells, as observed with PI uptake by the nonviable cells, was not energy dependent, indicating that the interaction does not require a cellular metabolic function. It was thus suggested that the fungicidal activity of indolicidin was mediated by a cellular function not requiring cellular energy consumption. Recently, it has been proposed that the ion channel raises the concentration of salt on the epithelial surface and this high salt concentration inhibits the activity of antimicrobial peptides. In order to evaluate whether salt ions affect the fungicidal activity of synthetic peptides, the effects of sodium chloride and magnesium chloride on the fungicidal effect of indolicidin were investigated using 1,6-diphenyl-1,3,5-hexatriene (DPH) as a membrane probe. If the fungicidal activity of indolicidin against *T. beigelii* is inhibited by salts, DPH, that interacts with an acyl group of plasma membrane lipid bilayers, could not be inhibited for insertion into the membrane. The fungicidal activity of the peptide increased plasma membrane DPH fluorescence anisotropy by adding salts. At 20 μM indolicidin, the presence at 20mM NaCl and $MgCl_2$ caused an inhibition of about 95% in the extent of permeabilization. This inhibition likely depends on the stabilizing effect of Na^{2+} and Mg^{2+} on fungal membranes through interaction with anionicsit es on membrane lipid components. The initial binding of indolicidin is thought to depend on a membrane interaction that is vulnerable to increase in salt concentrations and is followed by the formation of ion channels. This result has demonstrated that indolicidin possesses a fungicidal effect, which is affected by the electrostatic interaction and depends heavily on the ionic strength.

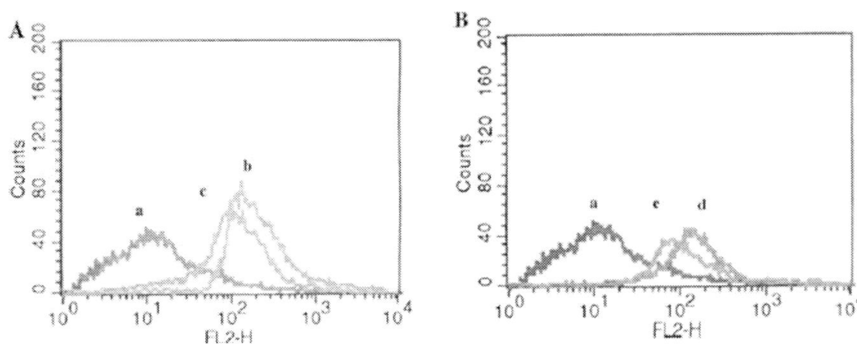

Figure 3. The effect of sodium azide (NaN_3) on the indolicidin. Peptides were added to a final concentration of 20 μM and the 0.05% NaN_3 as a respiration inhibitor, were added for energy-dependent test. T. beigelii cells were incubated in the absence (a) or presence (b) of indolicidin with 0.05% NaN_3 (c), or the presence (d) of melittin with NaN_3 (e) [18].

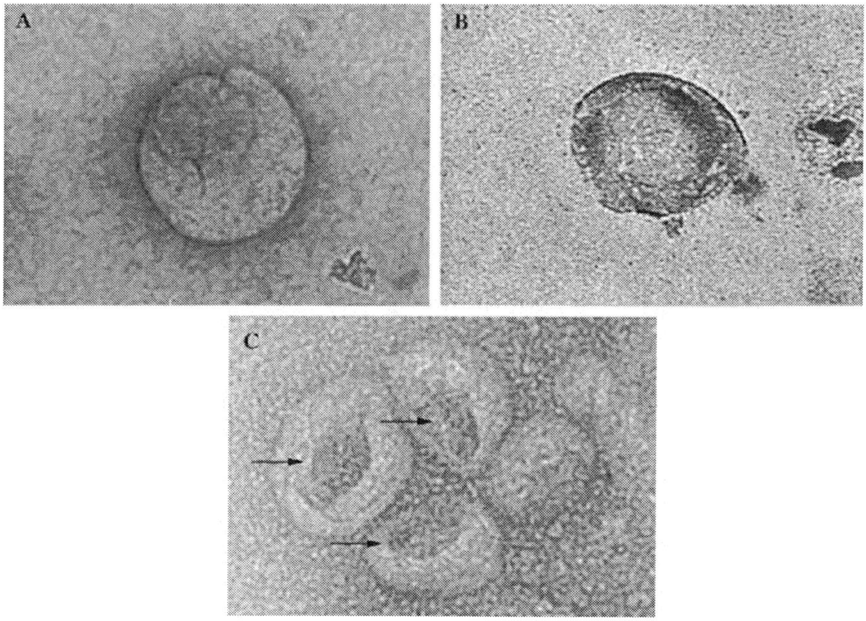

Figure 4. Electron micrographs of negatively stained SUV composed off PC/cholesterol (10:1. w/w) in the absence (A), or in the presence (B), of indolicidin, or melittin (C) [18].

In order to confirm the ability of indolicidin to disrupt microbial plasma membrane, an experiment was performed with liposomes. Artificial small unilamellar vesicles (SUVs) (PC/cholesterol; 10:1, w/w) and neutral vesicles of phosphatidylcholine (PC) were used as model membrane systems. The antimicrobial activity of melittin is based on the formation of transmembrane channels. Pores were formed from SUVs, after treatment with indolicidin or melittin (Figure 4), suggesting that the peptides may cause perturbation of the lipid components of the plasma membranes. It is conceivable that the effect of the peptide has the following stages: (i) binding to the membrane; (ii) membrane destabilization, and (iii) stabilization of the peptide–lipid complex. These results provide additional evidence that indolicidin probably acts on the plasma lipid membrane, by forming pores, causing the leakage of ions and other materials from the cells.

ARENICIN-1

Arenicin-1 (RWCVYAYVRVRGVLVRYRRCW), a 21-residue antibacterial peptide, is a β-sheet peptide isolated from the coelomocytes of the marine polychaeta lugworm, Arenicola marina. This peptide has been reported to exhibit broad antibacterial activity against human pathogenic bacterial strains. The structure of arenicin-1 was found to consist of two stranded anti-parallel β-sheet, constrained by one disulfide bond and connected by a β-turn. As arenicin-1 bears a net positive charge (+6) in combination with a substantial number of hydrophobic amino acid residues, extensive research regarding a membranolytic steps involved in the killing of bacterial strains have been conducted. However, its mechanism is still far from being fully answered. For this reason, we had investigated three different kinds of studies about antimicrobial effects of arenicine-1 and how does Trp residue affect its antimicrobial effects [19-21]. To determine the Trp requirements of arenicin-1 in exerting antimicrobial activity, a truncated peptide with an N-terminal deletion (analogue 1) was characterized by comparison to arenicin-1 (Table 2).

Table 2. Amino acid sequences of the synthetic antimicrobial peptide, arenicin-1 and its analogue [20]

Peptides	Amino acid sequences	Net charge	Retention time (min)	Remarks
arenicin-1	RWCVYAYVRVRGVLVRYRRCW	+6	20.133	wild-type peptide
analogue 1	CVYAYVRVRGVLVRYRRCW	+5	18.933	arenicin-1:N-terminal deletion (RW)

First of all, we investigated the activity of arenicin-1 and its analogue 1 on human pathogenic microorganisms. In this study, melittin was used as a positive control that has a strong cytolytic potency in both eukaryotic and prokaryotic cells. The results showed that arenicin-1 and the analog possess an antimicrobial effect, but exhibited less potent antimicrobial activity than melittin. Furthermore, its analog was also less potent than arenicin-1 (Table 3).

Table 3. The antimicrobial activity of arenicin -1and its analogue [19-21]

Bacterial strains	MIC (μM)		
	arenicin-1	analogue 1	Melittin
Gram-positive bacteria			
E. faecium ATCC29212	4.5	9	0.56
VRE[a]	2.25	4.5-9	1.125
S. aureus ATCC25923	2.25	9	1.125
MRSA[b]	2.25-4.5	9	1.125
Gram-negative bacteria			
E. coli ATCC25922	2.25	4.5-9	1.125
AREC[c]	2.25-4.5	4.5	2.25
P. aeruginosa ATCC27853	2.25-4.5	9	1.125
ARPA[d]	2.25	9	1.125
Fungi			
C. albicans ATCC 90028	9	36	2.25
M. furfur KCTC 7744	9	18	2.25
T. beigelii KCTC 7707	4.5	9	1.625
T. rubrum KCTC 6345	9	18	2.25

[a] *Vancomycin-Resistant E. Faecium*
[b] *Methicillin-Resistant S. aureus*
[c] *Antibiotics-Resistant E. coli*
[d] *Antibiotics-Resistant P. aeruginosa*

The hydrophobicity value of the antimicrobial peptides influences the interaction between the peptides and cell membranes. In human cell membranes, the membrane consists of neutral sterols including cholesterols and ergosterols, and most anionic phospholipids are gathered in the intra cytoplasmic leaflet. Consequently, the degree of cationic propensity is not considered in the interaction between the membranes and peptides. Based on the fact that hemolytic activity is related to the hydrophobicity of peptides, the hemolytic activity on human red blood cells was measured at various peptide concentration levels (Table 4). The analog exhibited a less potent hemolytic activity than arenicin-1 on mammalian erythrocytes at all concentrations.

Table 4. Hemolytic activity of arenicin-1 and its analogue against human erythrocyte cells [21]

Peptides	% Hemolysis (μM)					
	100	50	25	12.5	6.25	3.12
arenicin-1	51	35	12	0	0	0
analogue-1	4	0	0	0	0	0

In particular, analogue 1 revealed no hemolytic activity except at the highest concentration. It is assumed that the removal of tryptophan with its hydrophobic characteristics reduced hemolysis. These results show that the hydrophobicity of the analogue 1 was reduced.

Now, we will disclose the mechanisms of their antifungal- or antibacterial activity separately. A suggestion is thought that membranolytic antifungal activity of arenicin-1 requires the N-terminal tryptophan. Arenicin-1 exhibits membrane-disruptive mechanisms against *Candida albicans*. Therefore, to confirm the mechanisms and evaluate the membrane defects caused by arenicin-1 and its analogue, the release of FITC-dextran from liposomes was investigated. This method can also be employed to confirm the poreforming action of peptides. *Candida albicans* membrane mimetic liposomes composed of PC:PE: PI:ergosterol (5:4:1:2, w/w/w/w) and containing FITC-dextran (MW 3,380, 10,100 and 14,800) were prepared, and the release of entrapped dye from the vesicles was monitored as fluorescence level. As shown in Figure 5, the membrane-disruptive capability of the arenicin-1 and its analogue was consistent with the results of the MIC tests, possibly because of changes in the physicochemical properties of the analogue. To further elucidate the sitespecific role of the N-terminal Trp on arenicin-1 antifungal activity, peptide–lipid interactions in the lipid phase were investigated.

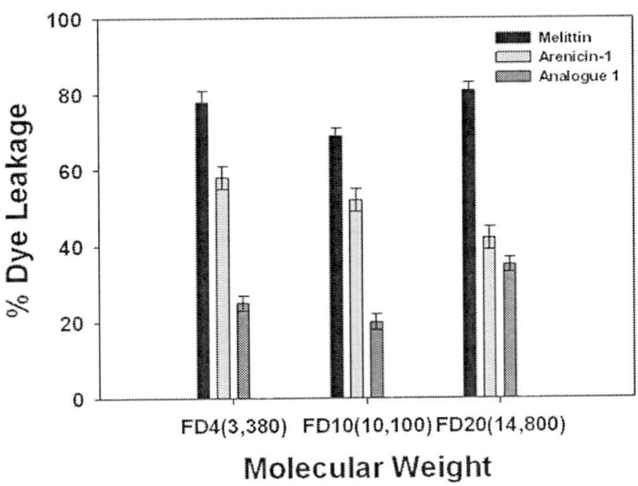

Figure 5. FITC-dextran leakage from C. albicans mimetic liposomes. C. albicans membrane mimetic liposomes composed of PC:PE:PI: ergosterol (5:4:1:2. w/w/w/w) containing FITC-dextrans of FD4 (3,380 Da), FD10 (10,100 Da), FD20 (14,800 Da) were incubated with peptides. Values are the mean of three independent measurements [20].

Table 5. Tryptophan emission maxima of the peptides in 5mH HEPES, 0.1mM EDTA buffer (pH 7.0), or in the presence of PC:PE:PI: ergosterol (5:4:1:2, w/w/w/w) SUVs [20]

Peptides	5mM HEPES, 0.1mM EDTA buffer	PC:PE:PI:ergosterol
Arenicin-1	350	339 [a]
Analogue 1	357	354
Melittin	349	336

[a] Blue shift in emission maximum in parentheses

The sensitivity of Trp fluorescence emission to its environment allowed us to monitor peptide binding to cell membranes. As the polarity of the environment decreases, the wavelength of Trp fluorescence becomes shorter (blue shift). Table 5 shows that the peptides in distilled water had a fluorescence emission maximum of 349–357 nm. Upon the addition of *C. albicans* membrane mimetic liposomes, a blue shift in the fluorescence emission maximum of Trp (3–13 nm) was observed. Arenicin-1 showed a blue shift of 11 nm, analogue 1 shifted by 3 nm. These results indicated that the peptides penetrated the hydrocarbon region of the bilayer, and based on the blue shift, the depth of penetration of each peptide was compatible to its activity. Since analogue 1 showed the most attenuated activity, the N-terminal Trp was likely involved in membrane penetration associated with disruption of the lipid bilayer. We investigated the structural requirements for the antifungal activity of the beta hairpin antimicrobial peptide arenicin-1. By comparison to the wild type peptide, the designed analogue showed a reduction in membrane-disrupting activity, indicating that the N-terminal Trp was involved in the penetration of peptides into the lipid bilayer. In the minimum inhibitory concentration (MIC) test, the antibacterial activities of the analogue 1 was reduced for both Gram-positive and Gram-negative bacteria, when compared to arenicin-1. These results suggest that the Trp residue in the N-terminal region of arenicin-1 plays a key role in antibacterial activity. It is known that arenicin-1 exerts its antibacterial effect by disrupting the plasma membrane. To compare the effect of arenicin-1 and its analogue on the integrity of bacterial membranes, a change in fluorescence intensity induced by the peptides was investigated by monitoring the PI influx. PI, a membrane-impermeant dye, only enters cells that have damaged membranes, after which the fluorescence of the probe is enhanced 20 to 30-fold. If the cell membranes were disrupted by the peptides, PI could then permeate into the cytoplasm and bind to the DNA. The PI influx of the analogue 1 was less potent than that of arenicin-1 in both Gram-positive and Gram-negative bacteria. Using DPH as membrane probe, the effect of arenicin-1 and its analogue on the bacterial membrane was investigated. DPH enables it to interact with an acyl group of the lipid bilayer within the plasma membrane without disturbing the structure of the lipid bilayer. If the plasma membrane is damaged by the antimicrobial peptides, then the DPH molecules cannot be inserted into the lipophilic tails of the phospholipids in the bilayer. This allows for a comparison of the antibacterial activities of arenincin-1 and its analogue on cell membranes and their potencies (Figure 6). The DPH fluorescence intensity of the analogs decreased when increasing their concentrations, but this decrease was still less than that observed in the presence of melittin and arenicin-1. Analogue 1 showed a lower decrease in the fluorescence intensity in Gram-negative bacteria and Gram-positive bacteria. It seems to be deprived of the capability to penetrate the outer membrane and cell wall. As the Trp residue of the peptide is sensitive to a hydrophobic environment, this sensitivity provided a possibility for monitoring its binding to the liposome. Trp residue in HEPES/EDTA buffer has an emission maximum by excitation at 280 nm and an enhanced emission peak centered at 350 nm. We used liposomes based on the following lipid mixtures that mimic the membrane compositions of bacteria: *S. aureus* (PG:CL = 58:42, w/w) and *E. coli* (PG:PE = 3:1, w/w). When the peptides bind to liposomes that mimic Gram-positive or Gram-negative bacterial membranes, the emission maximum of the peptides having the Trp residues will shift to a shorter wavelength (blue shift). While arenicin-1 and melittin showed a significant blue shift in liposomes, the analogue 1 had a lower blue shift (Table 6). These results suggest that the analogue 1 has less binding affinity to the bacterial membrane and have difficulty in deeply penetrating the hydrocarbon region of the anionic liposomal bilayer. To investigate the membrane permeating ability of arenicin-1 and its analogue, the release of fluorescently-labeled dextran molecules of various sizes (FD4, 3.9 kDa, 1.8 nm radii; FD10, 9.9 kDa,2.3 nm radii; FD20, 19.8 kDa, 3.3 nm radii) was measured. The FD-loaded liposomes were incubated with 10 μM of the peptides. Negatively charged LUVs were comprised of PG:CL (58:42, w/w) and PG:PE (3:1, w/w) to mimic microbial cells (Figure 7). The results show that arenicin-1 and its analogue bind to and make pores in the bacterial cell membrane. As shown in Figure 7, the entrapped FD20 was released for all the peptides; this indicates that the radii of the pores created by the peptides were larger than 3.3 nm. The leakage of FD from liposomes treated with the analogue was less than that of arenicin-1. This implies that a decrease in the electrostatic interactions did not form enough pores to exert antibiotic activity. In conclusion, it was revealed that the analogue 1 with a truncated N-terminal region (RW) decreased net hydrophobicity and charge. It suggests that the Trp residues in the N-terminal region of arenicin-1 play a key role in its antibacterial activity.

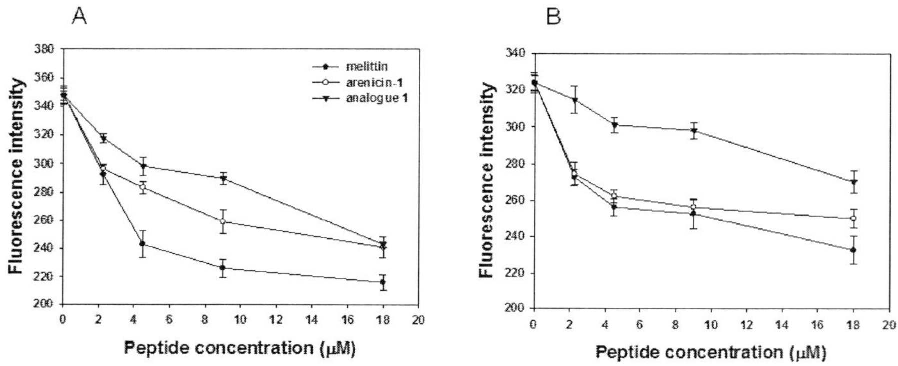

Figure 6. DPH fluorescence intensity after the addition of peptides. Bacterial cells containing the peptides of various concentrations (2.25, 4.5, 9, or 18 μM) were incubated. (A) *S. aureus* cells treated wite peptidesm and (B) *E. coli* cells treated with peptides [21].

Table 6. Tryptophan emission maxima of the peptides in 5mM HEPES, 0,1mM EDTA buffer (pH 7.0), or in the presence of PG:CL (58:42, w/w) and PG:PE (3:1. w/w) SUVs [21]

Peptides	buffer	PG:CL (58:42, w/w)	PG:PE (3:1, w/w)
arenicin-1	351	341 (9)[a]	339 (10)
analogue 1	345	342 (7)	341 (6)
melittin	348	335(12)	333(13)

[a] Blue shift in emission maximum in parentheses

Table 7. Amino acid sequence and molecular masses determined by MALDI-MS of CA-MA hybrid peptide and its analogues [25]

Peptides	Amino acid sequence	Remarks	Calculated values	Observed values
CA-MA	KWKLFKKIGIGKF LHSAKKF-NH_2	Parents peptide	2402.48	2403
CA-MA analogue P5	KWKKLLKKPLLK KLLKKL-NH_2	CA-MA: substitution of GIG→Pro $L^4 \rightarrow K^4$, $F^5 \rightarrow L^5$, $K^6 \rightarrow L^6$, $I^8 \rightarrow K^8$, $K^{12} \rightarrow L^{12}$, $F^{13} \rightarrow L^{13}$, $L^{14} \rightarrow K^{14}$, $H^{15} \rightarrow K^{15}$, $S^{16} \rightarrow L^{16}$, $A^{17} \rightarrow L^{17}$, $F^{20} \rightarrow L^{20}$	2244.6	2246
P6	KWKLKPLLKKLLK KL-NH_2	P5: *N*-terminal deletion (K^4, L^6, K^8)	1875.34	1876
P7	KWKKLLKKPLKL KL-NH_2	P5: *C*-terminal deletion (L^{11}, K^{13}, L^{15}, K^{17})	1762.24	1763
P8	KPLLKKLLKKL-NH_2	P6: *N*-terminal deletion (KWKL)	1320.9	1320
P9	KLLKKPLKLKL-NH_2	P5: *N*-terminal deletion (KWK)	1320.9	1322

Peptides were synthesized by the solid phase method using Fmoc(9-fluorenyl-methoxycarbonyl)-chemistry. The molecular masses of the peptides were confirmed with a MALDI (matrix-assisted laser desorption/ionization) mass spectrometer.

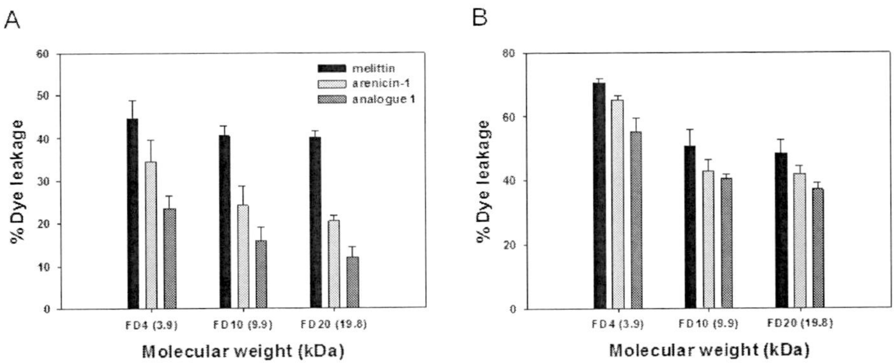

Fugure 7. Percentage of FD leakage induces by the peptides. (A) LUVs composed of PG:CL (58:42. w/w, and (B) LUVs composed of PG:PE (3:1. w/w) liposome [21].

Cecropin

Cecropin A, one of the first reported antimicrobial peptides, is found in the hemolymph of *Hyalophora cecropia* pupae and consists of 37 amino acid residues. Magainin 2, a cationic 23-amino acid antimicrobial peptide, was discovered in the skin of the African clawed frog, *Xenopus laevis* [22]. Both cecropin A (CA) and magainin 2 (MA) exhibit strong antibacterial activity but no cytotoxicity against normal mammalian cells [23]. Numerous studies using synthetic peptides have been focused on designing analogue peptides with increased antimicrobial activity compared to that of natural peptides without damaging the mammalian cells [24]. In previous study, several novel analogue peptides (P6, P7, P8 and P9) with antimicrobial activity were designed by chain length deletion and synthesized from the sequence of the CA-MA analogue, P5 [25, 26]. To investigate the correlation between antibiotic activity and the analogue peptides, the antimicrobial and anticancer activity were measured against bacterial, fungal and cancer cells. Additionally, we will discuss the effect of analogue peptides on the cell membrane.

The amino acid sequences used in this study were summarized and the correct molecular weights of the synthetic peptides were confirmed by MALDI mass spectrometry (Table 7). The several analogues were designed to change the net positive charge and hydrophobicity by Lys- and Leu-deletion or *N*-terminal region (KWK or KWKKLLK)-deletion. In particular, it was shown that the tryptophan residue in position 2 of the P6 was critical in antibacterial activity [27].

Antibacterial activities of the synthetic peptides against Gram-positive and Gram-negative bacterial strains were determined as the MIC (Table 8). Tryptophan residue at position 2 (P6) showed strong antibacterial activity than other analogue peptides with a MIC value of 0.19–3.125 μM. Both *N*-terminal region (KWK or KWKKLLK)-deletion and the tryptophan residue deletion at position 2 caused a 2 or 4-fold decrease in the antibacterial activity against all the bacterial cells tested. The analogue P7 (L:K, 5:7) was similar to or half as active as P6 (L:K, 6:7).

This result indicates that the moderate chain-length, net positive charge and hydrophobicity of synthetic peptide are important in maintaining the antibacterial activity. The antifungal activity of the peptides against the pathogenic fungi was measured as MIC by MTT assay. The results indicated that P6 displayed a 4-fold increased antifungal activity than other analogue peptides against *T. beigelii* (Table 8), which is similar to antibacterial activity. The results thought it possible that tryptophan residue is a crucial part of the antimicrobial activity of the analogue peptides used in this study.

Table 8. Antimicrobial activity of CA-MA hybrid peptide and its analogues [25,26]

Peptides	MIC (μM)						
	Gram-positive bacteria		Gram-negative bacteria		Fungal strains		
	B. subtilis	*S. epidermidis*	*P. aeruginosa*	*S. typhimurium*	*C. albicans*	*T. beigelii*	*S. cerevisae*
CA-MA	3.1	1.6–3.1	3.1	0.2	12.5	1.6–3.1	12.5
CA-MA analogue P5	1.6	1.6	1.6	0.2	6.3	1.6	6.3
P6	3.1	1.6	3.1	0.2	12.5	1.6	12.5
P7	3.1	3.1	3.1	0.2	12.5	3.1	12.5
P8	6.3	6.3	3.1	0.4	25	6.3	12.5
P9	12.5	6.3	6.3	0.8	>25	6.3	>25

The bacteria were grown to the mid-growth phase in a medium (g l^{-1}) (10 bactotryptone/5 yeast extract/10 NaCl (pH 7)). Microbial growth was determined by the increase in optical density at 620 nm after 10 h incubation at 37 °C. The fungal strains were grown at 30 °C in YPD media (Dextrose 2%, Peptone 1%, yeast extract 0.5%, pH 5.5). The fungal cells were seeded on the well of a 96-microtiter plate of YPD media at a density of 2×10^3 cells (100 μl per well). The turbidity of each well was measured at 580 nm using a microtiter ELISA reader (Molecular Devices Emax, CA).

INDOLE-3-CARBINOL

Tryptophan has a indole sidechain, Indole-3-carbinol is based on indole sidechain (Figure 8). Indole-3-carbinol (I3C) is a compound found in high concentrations in *Brassica* vegetables, including broccoli, cauliflower, *Brussels sprouts*, and cabbage. It is produced from naturally occurring glucosinolates contained in a wide variety of plants. Because of its anticarcinogenic effects in experimental animals and humans, I3C has received special attention as a possible chemopreventive agent, I3C has also been found to inhibit the growth of various cancer cells, including those of the breast and prostate. Although there are some conflicting reports in the literature, such as the exposure of I3C during the postinitiation phase being shown to have carcinogenic effects, numerous reports favoring the anticarcinogenic effects of I3C are overwhelming. We had reported on the antimicrobial effects of I3C against various human pathogens and its mode of action regarding antifungal activity against fungal pathogens [28]. The antimicrobial effects of I3C on human pathogenic microorganisms, including the clinical isolates of antibiotic-resistant bacteria, were investigated and described as MIC [29].

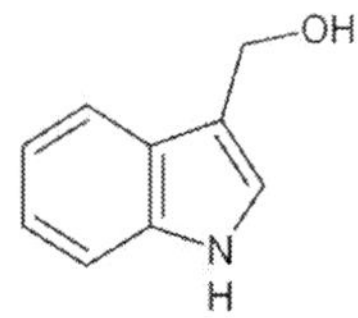

Figure 8. Chemical Structure of I3C.

Table 9. Antimicrobial Activity of Indole-3-Carbinol against Human Pathogenic Microorganisms [28, 29]

Microorganisms	MIC (μg/ml)		
	Indole-3-carbinol	Propionic acid	Amphotericin B
Gram-positive bacteria			
S. aureus (wild type)	5	20	NT[f)]
S. epidermidis	5	40	NT
MRSA 1[a)]	5	40	NT
MRSA 2	10	40	NT
MRSA 3	5	20	NT
MRSA 4	5	20	NT
E. faecium (wild type)	20	40	NT
VSE[b)]	20	80	NT
VRE 1[c)]	10	120	NT
VRE 2	20	80	NT
VRE 3	20	120	NT
Gram-negative bacteria			
E. coli (wild type)	40	20	NT
E. coli O-157	40	10	NT
MREC 1[d)]	40	20	NT
MREC 2	80	20	NT
MREC 3	40	20	NT
P. aeruginosa (wild type)	40	40	NT
MRPA 1[e)]	40	40	NT
MRPA 2	20	40	NT
MRPA 3	40	40	NT
MRPA 4	20	40	NT
Fungi			
C. albicans	10	NT	5
T. beigelii	20	NT	2.5
S. cerevisiae	10	NT	5

a) Methicillin-resistant Staphylococcus aureus. b) Vancomycin-sensitive Enterococci. c) Vancomycin-resistant Enterococci. d) Multiantibiotic-resistant Escherichia coli. e) Multiantibiotic-resistant Pseudomonas aeruginosa. f) Not tested.

In this study, propionic acid and amphotericin B were used as a positive control toward bacteria and fungi, respectively; propionic acid is an antibacterial agent widely used as a food preservative, and amphotericin B is a fungicidal agent widely used in treating serious systemic infections. I3C, in an MIC value of 5—20μg/ml, showed antibacterial activity against gram-positive bacterial strains including the clinical isolates of methicillin-resistant *Staphylococcus aureus* (MRSA), and vancomycin-sensitive or resistant Enterococci (VSE or VRE). I3C exhibited more potent activity than propionic acid, showing MIC values of 20—120μg/ml on gram-positive bacterial strains (Table 9). In a separate study, I3C, in an MIC value of 20—80μg/ml, showed antibacterial activity against gram-negative bacterial strains including the clinical isolates of multiantibiotic-resistant *Escherichia coli* (MREC), and multiantibiotic-resistant *Pseudomonas aeruginosa* (MRPA). I3C exhibited less potent activity than propionic acid, showing MIC values of 10—40μg/ml on gram-negative bacterial strains (Table 9). In testing fungi, I3C also showed antifungal activity against human pathogenic fungal strains. This compound exhibited MIC values within a concentration range of 10—20μg/ml, and the MIC values of this compound were as high as those of amphotericin B, showing MIC values of 2.5—5μg/ml on all fungal strains (Table 9). To understand how the activity of I3C affects cellular physiology, we investigated the effects on the cell cycle progress of *C. albicans*. The cells were cultured for 6 h at 28 °C in the presence or absence of I3C, and the DNA content was determined *via* flow cytometry by staining with PI. PI is a DNA-staining dye that intercalates between the bases of DNA or RNA molecules.

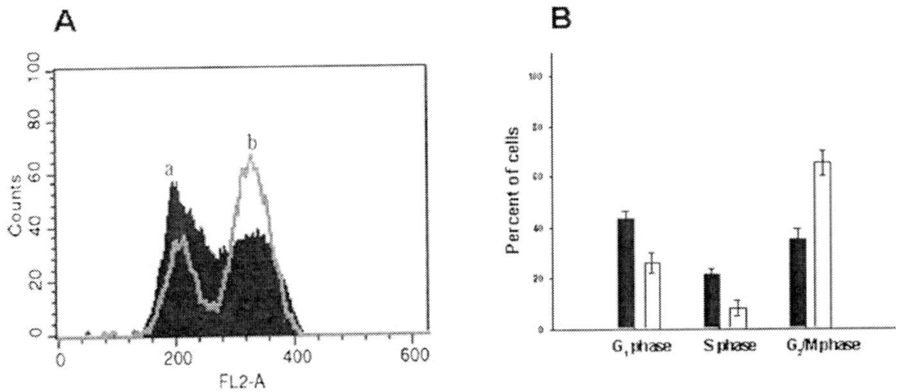

Figure 9. The Effects of Indole-3-Carbinol (I3C) on the Process of Cell Cycle of C. albicans, C. albicans cells were treated with 100m g/m of I3C and incubated at 28 °C for 6 h under constant shaking. After washing the cells with a PBS, the cells were fixed in 70% ethanol (in PBS, v/v) for 12 h, and then, stained with 50m g/ml of propidium iodide. (A) Cell cycle histogram of C. albicans, (B) graphical presentation of the percentage of cell cycle in C. albicans: a and (black bars), not treated with any compound: b and (white bars), treated with I#C, and the errot bars represent the standard deviation (S.D.) values for three independent experiments, performed in triplicate [28].

As shown in Figure 9, the percentage of cells in the G2/M phase increased by approximately 30%, while that in the G1 phase significantly decreased by about 17%, and that in the S phase also decreased by about 13% in the presence of I3C (Figure 9).

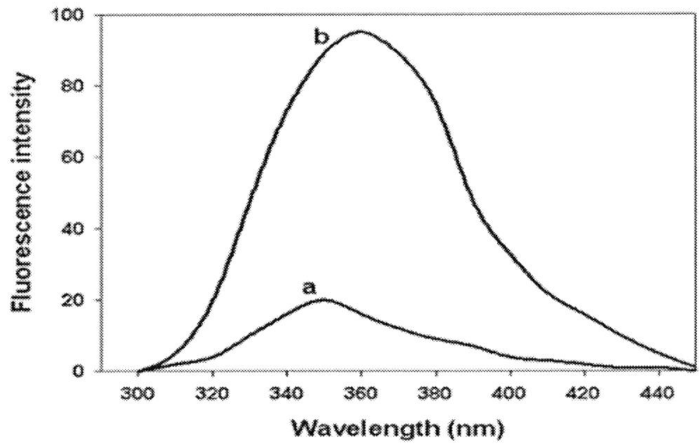

Figure 10. Fluorescence Emission Spectrum of Indole-3-Carbinol in the Presence (a) or Absence of Liposomes (b) [28].

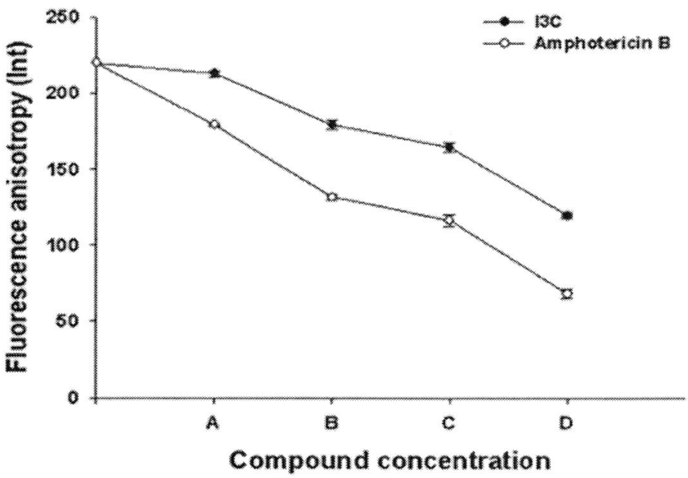

Figure 11. DPH Fluorescence Anisotropy after the Addition of Indole-3-Carbinol (I3C) and Amphoterican B. Treatment with A. 40: B. 80: C. 120: D. 160m g/ml of I#C or treatment with A. 20: B. 40: C. 60: D. 80 μg/ml of amphotericin B. and the error bars represents the standard deviation (S.S.) values for three independent experiments, performed in triplicate [28].

To examine the interaction of the I3C with the phospholipids membrane, we measured its fluorescence emission spectra in HEPES buffer and in the presence of PC/cholesterol (10 : 1, w/w) SUVs. The indole ring of I3C provided the possibility to monitor its binding to liposome since the fluorescence emission of this ring is sensitive to the polarity of its environment. In HEPES buffer, I3C displayed fluorescence emission maximums at 360 nm, which is typical for indole ring in polar surroundings, suggesting the indole ring of I3C was located in a hydrophilic environment. The addition of PC/cholesterol (10 : 1, w/w) vesicles caused a significant blue shift (approximately 10 nm) in the emission maximum (Figure 10).

The effect of I3C on the fungal plasma membrane was further investigated by using DPH as a membrane probe. If the antifungal activities exerted by I3C on *C. albicans* were at the level of the plasma membrane, DPH, which interacts with an acyl group of the plasma membrane lipid bilayer, could not be inserted into the membrane. As shown in Figure 11, the plasma membrane DPH fluorescence anisotropy was significantly decreased by increasing the concentrations of I3C and amphotericin B. This was consistent with the disruption of the plasma membrane by I3C as well as by a positive control, amphotericin B.

In summary, I3C exhibited potent antimicrobial effects on various pathogenic microorganisms. Although the exact mechanism of the I3C action has not yet been fully elucidated, the results reported here indicate its effect on the plasma membrane.

CONCLUSIONS

The Trp residue may play an important role as antimicrobial peptides and compounds are being tailored for clinical use. Arenicin-1 (RWCVYAYVRVRGVLVRYRRCW), a 21-residue antibacterial peptide, is a β-sheet peptide isolated from the coelomocytes of the marine polychaeta lugworm, Arenicola marina. Indolicidin (ILPWKWPWWPWRR), from cytoplasmic granules of bovine neutrophils, has a unique composition consisting of 39% tryptophan. Indole-3-carbinol is based on indole sidechain of tryptophan. Interestinly, these peptides or compound have antimicrobial activities against broad range of microorganisms in common. Especially, the study about comparing with arenicin-1 and its analogue had shown that the antimicrobial activities of the analogue were lower than arenicin-1. And aside from this, we had studied of antimicrobial effects of other substances, including pleurocidin and melittin [30]. The common denominator within these substances is that all they have tryptophan residue and also possess antimicrobial activities. Subsequently, the Trp play a role in membrane association because they preferably interact with the interfacial region of membranes. Similarly, this can be applied to antimicrobial peptides, where in aqueous solution, the Trp residues help maintain a few hydrophobic contacts which retain the structure in a conformation favorable for subsequent membrane interactions. Overall, because of their many advantageous properties, it appears that Trp residue is exceedingly suitable to contribute to potent antimicrobial agents that eventually may be used on humans.

Further Reading

[1] M. Zasloff, Magainins, a class of antimicrobial peptides from Xenopus skin-isolation, characterization of two active forms, and partial cDNA sequence of a precursor, *Proc. Natl. Acad. Sci. U. S. A.* 84 (1987) 5449–5453.

[2] J.H. Andersen, H. Jenssen, K. Sandvik, T.J. Gutteberg, Anti-HSVactivity of lactoferrin and lactoferricin is dependent on the presence of heparin sulphate at the cell surface, *J. Med. Virol.* 74 (2004) 262–271.

[3] A.J. De Lucca, T.J. Walsh, Antifungal peptides: novel therapeutic compounds against emerging pathogens, *Antimicrob. Agents Chemother.* 43 (1999) 1–11.

[4] N. Papo, Y. Shai, Host defense peptides as new weapons in cancer treatment, *Cell. Mol. Life Sci.* 62 (2005) 784–790.

[5] R. Jerala, M. Porro, Endotoxin neutralizing peptides, *Curr. Top. Med. Chem.* 4 (2004) 1173–1184.

[6] I. C. David, J. P. Elmar, J. V Hans, Tryptophan- and arginine-rich antimicrobial peptides: Structures and mechanisms of action, *Biochimica et Biophysica Acta.* 1758 (2006) 1184-1202.

[7] K.A. Brogden, Antimicrobial peptides: pore formers or metabolic inhibitors in bacteria? *Nat. Rev., Microbiol.* 3 (2005) 238–250.

[8] W.M. Yau, W.C. Wimley, K. Gawrisch, S.H. White, The preference of tryptophan for membrane interfaces, *Biochemistry.* 37 (1998) 14713–14718.

[9] S. Persson, J.A. Killian, G. Lindblom, Molecular ordering of interfacially localized tryptophan analogs in ester- and ether-lipid bi-layers studied by H-2-NMR, *Biophys. J.* 75 (1998) 1365–1371.

[10] J.A. Killian, I. Salemink, M.R.R. de Planque, G. Lindblom, R.E. Koeppe, D.V. Greathouse, Induction of nonbilayer structures in diacylphosphatidylcholine model membranes by transmembrane alpha-helical peptides: Importance of hydrophobic mismatch and proposed role of tryptophans, *Biochemistry*. 35 (1996) 1037–1045.
[11] J. Klein-Seetharaman, M. Oikawa, S.B. Grimshaw, J. Wirmer, E. Duchardt, T. Ueda, T. Imoto, L.J. Smith, C.M. Dobson, H. Schwalbe, Long-range interactions within a nonnative protein, *Science*. 295 (2002) 1719–1722.
[12] D.A. Dougherty, Cation–pi interactions in chemistry and biology: a new view of benzene Phe, Tyr, and Trp, *Science*. 271 (1996) 163–168.
[13] J.C. Ma, D.A. Dougherty, The cation–pi interaction, *Chem. Rev*. 97 (1997) 1303–1324.
[14] M.P. Aliste, J.L. MacCallum, D.P. Tieleman, Molecular dynamics simulations of pentapeptides at interfaces: salt bridge and cation–pi interactions, *Biochemistry*. 42 (2003) 8976–8987.
[15] F.N. Petersen, M.O. Jensen, C.H. Nielsen, Interfacial tryptophan residues: a role for the cation–pi effect? *Biophys. J.* 89 (2005) 3985–3996.
[16] H. G. Boman, Peptide antibiotics and their role in innate immunity, *Annu. Rev. Imm*unol. 13 (1995) 61-92.
[17] R. E. V. Hancock, Peptide antibiotics, *Lancet*. 349 (1997) 418-422.
[18] D. G. Lee, H. K. Kim, S. A. Kim, Y. Park, S. C. Park, S. H. Jang, K. S. Hahm, Fungicidal effect of indolicidin and its interaction with phospholipid membranes, *Biochem. Biophys. Res. Commun*. 305 (2003) 305-310.
[19] C. Park, D. G. Lee, Fungicidal effect of antimicrobial peptide arenicin-1, Biochimica. et. *Biophysica. Acta*. 1788 (2009) 1790-1796.
[20] C. Park, J. Cho, J. Lee, D. G. Lee, Membranolytic antifungal activity of arenicin-1 requires the N-terminal tryptophan and the beta-turn arginine, *Biotechnol. Lett*. 33 (2011) 185-189.
[21] J. Cho, D. G. Lee, The characteristic region of arenicin-1 involved with a bacterial membrane targeting mechanism, *Biochem. Biophys. Res. Commun*. 405 (2011) 422-427.
[22] M. Zasloff, Magainins, A class antimicrobial peptides from *Xenopus* skin: isolation, characterization of two active forms, and partial cDNA sequence of a precursor, *Proc. Natl. Acad. Sci. USA*. 84 (1987) 5449-5453.

[23] H. K. Kim, D. G. Lee, Y. Park, H. N. Kim, B. H. Choi, C. H. Choi, K. S. Hahm, Antibacterial activities of peptides designed as hybrids of antimicrobial peptides, *Biotechnol. Lett.* 24 (2002) 347-353.

[24] M. M. Javadpour, M. M. Juban, W. C. Lo, S. M. Bishop, J. B. Alberty, S. M. Cowell, C. L. Becker, M. L. McLaughlin, *De novo* antimicrobial peptides with low mammalian cell toxicity, *J. Med. Chem.* 39 (1996) 3107-3113.

[25] Y. Park, D. G. Lee, K. S. Hahm, Antibiotic activity of Leu-Lys rich model peptides, *Biotechnol. Lett.* 25 (2003) 1305-1310.

[26] Y. Park, D. G. Lee, S. H. Jang, E. R. Woo, H. G. Jeong, C. H. Choi, K. S. Hahm, A Leu-Lys rich antimicrobial peptide: activity and mechanism, *Biochim. Biophys. Acta.* 1645 (2003) 172-182.

[27] J. H. Kang, S. Y. Shin, S. Y. Jang, M. K. Lee, K. S. Hahm, Release of aqueous contents from phospholipid vesicles induced by cecropin A (1-8)-magainin 2 (1-12) hybrid and its analogues, *J. Peptide Res.* 52 (1998) 45-50.

[28] W. S. Sung, D. G. Lee, *In vitro* antimicrobial activity and the mode of action of indole-3-carbinol against human pathogenic microorganisms, *Biol. Pharm. Bull.* 30 (2007) 1865-1869.

[29] W. S. Sung, D. G. Lee, Mechanism of decreased susceptibility for Gram-negative bacteria and synergistic effect with ampicillin of indole-3-carbinol, *Biol. Pharm. Bull.* 31 (2008) 1798-1801.

[30] H. J. Jung, Y. Park, W. S. Sung, B. K. Suh, J. Lee, K. S. Hahm, D. G. Lee, Fungicidal effect of pleurocidin by membrane-active mechanism and design of enantiomeric analogue for proteolytic resistance, *Biochim. Biophys. Acta.* 1768 (2007) 1400-1405.

In: Tryptophan
Editors: B. L. Whitley et al.
ISBN 978-1-62100-445-5

Chapter IV

Metabolism of Tryptophan and Evaluation of Tryptophan Supplementation on Fish Larval Growth and Frequency of Skeletal Deformities

Margarida Saavedra[*]
Instituto Nacional de Investigação Agrária e das Pescas
(INIAP/IPIMAR-CRIPSul), Olhão, Portugal

Abstract

Amino acids are a major energy source during fish larval stage. Dietary amino acids imbalances have been described when fish larvae are fed rotifers, which may lead to a reduction in the growth rate. Amino acid imbalances may be overcome using amino acid supplementation. There are several ways to assess the effect of amino acid supplementation. The effect of amino acid supplementation can be tested at long term through zootechnique trials or at short term using the tube-feeding technique. Tryptophan is an indispensable amino acid and a precursor of the

[*]E-mail address: margarida.saavedra@gmail.com, Tel.: +351 289 71 53 46, Fax.: +351 289 71 55 79.

neurotransmitter serotonin, which affects food intake and aggression in fish. The levels of tryptophan have also been reported to be relevant to the incidence of fish skeletal malformations. Deformities in the vertebral column in reared fish are common and a constraint to the development certain fish cultures. This review studies tryptophan metabolism in a marine fish species, *Diplodus sargus* and evaluates the effects of tryptophan supplementation on fish larval survival, growth and incidence of skeletal deformities.

1. Introduction

One of the most serious problems aquaculture faces are the incidence of skeletal deformities. Fish skeletal deformities can have an important economical impact because they can affect fish size and shape, decreasing their economical value (Favaloro and Mazzola, 2000; Boglione et al., 2001). Most common malformations are vertebral deformities, which are often present in high frequencies in captive fish (Divanach et al., 1996). Vertebral malformations such as scoliosis, lordosis and kyphosis are often observed in fish (Kranenbarg et al., 2005) and caused by factors such as abnormal swim bladder insuflations (Chatain, 1994), dietary deficiencies (Lim and Lovell, 1978; Baeverfjord et al., 1998; Helland et al., 2005; Helland et al., 2006), high-current velocity (Divanach et al., 1997) and others. A high incidence of vertebral malformations may be associated to a poor hatchery performance, lower survival and growth rates and a higher susceptibility to stress and disease (Boglione et al., 2001). Throughout the years some of the causes for serious vertebral malformations have been considerably reduced with the development of new devices that, for example, reduce the amount of oil at the water surface, allowing a better swim bladder insuflation. In general, vertebral deformities are developed due to an insufficient knowledge of the optimum rearing techniques (Sfakianakis et al., 2005) and inadequate nutritional protocols (Cahu et al., 2003). Unbalanced amino acid (AA) diets seem to have an important impact on the frequency of vertebral deformities (Saavedra et al., 2009a). When given an AA balanced diet and an AA unbalanced diet to white seabream larvae, the ones fed the first diet showed almost half the frequency of skeletal deformities compared to larvae fed the AA unbalanced diet (Saavedra et al., 2009a). It has also been shown that in salmonids, scoliosis is often associated to tryptophan deficiency in the diet (Akiyama et al., 1986,

Hseu et al., 2003). Halver and Shanks (1960) also found high frequency of scoliosis and some lordosis in tryptophan-deficient sockeye salmon. Kloppel and Post (1975) reported considerable high frequencies of scoliosis in tryptophan-deficient rainbow trout. Tyryptophan seems to be, therefore, a relevant amino acid to prevent vertebral malformations in some fish species. The aim of this review is to study the metabolism of tryptophan in white seabream (*Diplodus sargus)*larvae and evaluate the effects of tryptophan supplementation on larval survival, growth and frequency of vertebral deformities.

2. Metabolism of Tryptophan

In this review tryptophan supplementation is analysed at two levels: a) using the *tube feeding* technique and b) through a zootechnique experimental trial. The first method determines the effect of tryptophan on its metabolism, AA catabolism and AA retention, and the zootechnique trial studies the effect of tryptophan supplementation on larval growth, survival and frequency of vertebral deformities.

Tube feeding was described by Rust et al. (1993) as an in vivo method, which allowed a controlled tube-feeding of fish larvae. This technique was later modified by Ronnestad et al. (2001) to be possible to estimate metabolic budgets of individual AA. *Tube feeding* was used to estimate the metabolic budgets of tryptophan in white seabream larvae (experimental set up described in details in Saavedra et al., 2008).

Four different treatments were used to study tryptophan metabolism: 1) Radiolabelled (^{14}C) tryptophan + supplement of crystalline tryptophan; 2) Radiolabelled (^{14}C) tryptophan + saline solution; 3) Mixture of several radiolabelled (^{14}C) AA + supplement of crystalline tryptophan; 4) Mixture of several radiolabelled (^{14}C) AA + saline solution. The first two treatments evaluate the efficiency of tryptophan supplementation and the comparison of the four treatments examines if tryptophan is limiting protein synthesis or not in this fish.

It was observed that larvae fed a tryptophan tracer showed lower AA retention compared to larvae fed the AA mixture tracer (Figure 1). On the other hand, larvae fed the tryptophan tracer had a higher percentage of evacuation compared to the average of all amino acids (Figure 1).

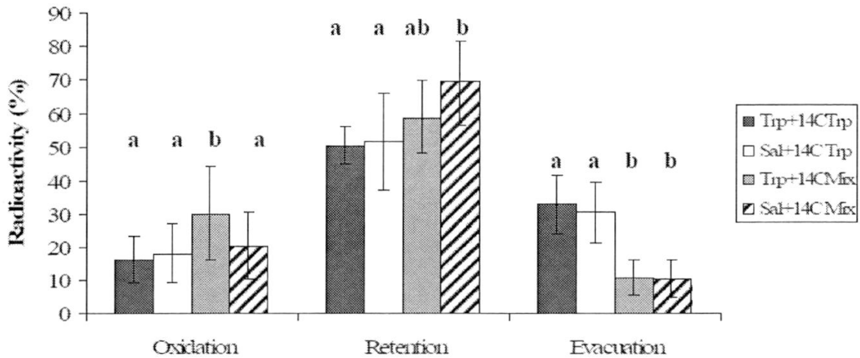

Figure 1. Percentage of the tube-fed ^{14}C amino acid mix (Mix) or ^{14}C tryptophan (Trp) oxidised, retained in body and evacuated by *Diplodus sargus* larvae. Trp+ ^{14}C Trp: Trp ^{14}C supplemented with crystalline tryptophan; Sal+ ^{14}C Trp: Trp ^{14}C without supplementation; Trp+ ^{14}C Mix: ^{14}C Mix supplemented with crystalline tryptophan ; Sal+ ^{14}C Mix: ^{14}C Mix without supplementation. Values are mean and standard deviation (n=10). Different letters correspond to significant differences ($p< 0.05$) (Data from Saavedra et al., 2008).

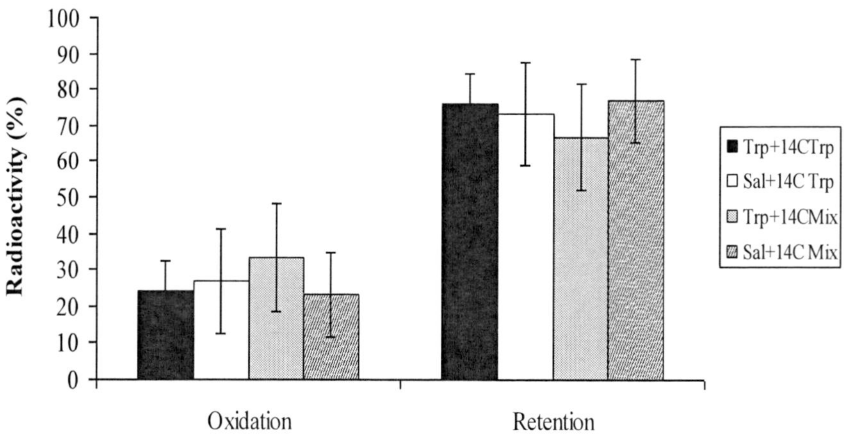

Figure 2. Percentage of oxidation and retention of ^{14}C AA in tube-fed *Diplodus sargus* larvae. Trp+ 14C Trp: Trp 14C supplemented with crystalline tryptophan; Sal+ 14C Trp: Trp 14C without supplementation; Trp+ 14C Mix: 14C Mix supplemented with crystalline tryptophan; Sal+ 14C Mix: 14C Mix without supplementation. Values are mean and standard deviation (Data from Saavedra et al., 2008).

Larvae fed the AA radioactive mixture and a supplement of tryptophan had significantly higher AA oxidation (Figure 2). These results suggest that tryptophan supplementation was efficient (Figure 2) and that this AA is not limiting *D. sargus* growth as there was no improvement in the retention percentage when a supplement of tryptophan was given to larvae tube-fed a labelled mixture of several AA (Figure 2).

3. Development of Skeletal Deformities During White Sea Bream Larval Ontogeny

In order to find the causes for fish skeletal deformities it is crucial to determine at which stage they start developing. Saavedra et al. (2010) run a trial to evaluate the frequency of vertebral malformations during D. *sargus* larval ontogeny. Newly hatched larvae from three different egg batches were examined during 30 days. Larvae were stock in 200 l tanks at a density of 80 larvae per tank. 30 larvae samples were taken from all tanks at 2, 3, 8, 13, 15, 18, 21, 23, 25, 27 and 30 days after hatched (DAH) to identify possible vertebral deformities. In order to identify the different structures, cartilage was stained with alcian blue for approximately 40 minutes and calcified structures (ossified bone and cartilage) was stained using alizarin red for 2 hours. This technique was done according to Gavaia et al. (2000). The bone and cartilage structures described by Koumoundouros et al. (2001) were used as a standard for this study. In this study the vertebral column was divided into three regions: trunk vertebrae (1 to 10), caudal vertebrae (11 to 20) and preurostyle vertebrae (21 to 23). The results showed that at early larval stages the frequency of skeletal deformities is very low and that is only when the larvae are older than 12 DAH that this frequency increases considerably. At 15 DAH there was already 80% showing at least one type of vertebral deformities. This percentage includes serious vertebral malformations such as kyphosis and scoliosis which was first observed when larvae were only 13 DAH. Preurostyle region was the region most affected by vertebral fusions and vertebral compresssions, occurring as early as 13 DAH. 40 to 60% of the larvae had vertebral compressions in this region of the vertebral column.

It was also observed abnormal shape vertebrae, mainly between 23 and 25 DAH, with a frequency of 20% in the caudal region and below 10% in the other two vertebral column regions. Hypertrophic vertebrae were observed in several larvae. These types of vertebrae were mainly found in the trunk region between vertebrae 5 and 12 and in early larval ages. After larvae were 21 DAH they were not observed any more.

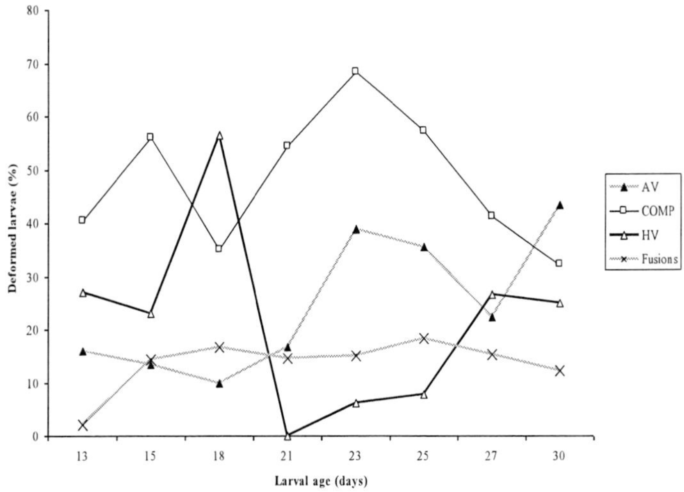

Figure 3. Percentage of skeletal deformities during *Diplodus sargus* larval ontogeny. AV- Abnormal vertebrae shape, COMP- vertebrae compression, HV- hypertrophic, F- Vertebrae fusion (data from Saavedra et al., 2010).

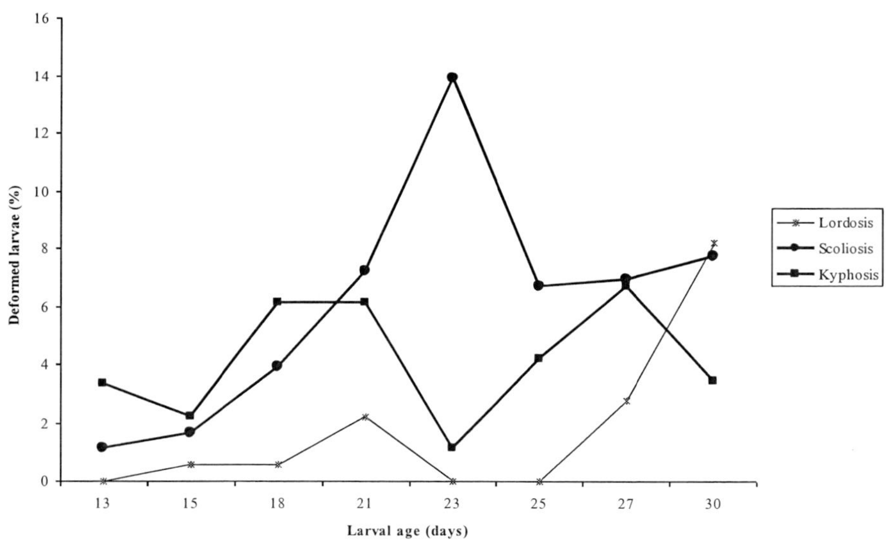

Figure 4. Percentage of *Diplodus sargus* larvae with lordosis, scoliosis and kyphosis during larval ontogeny (data from Saavedra et al., 2010).

4. SUPPLEMENTATION OF TRYPTOPHAN IN THE FISH LARVAL DIET

One of the main concerns about fish rearing is the high frequency of skeletal deformities observed. The presence of a significant number of fish with these deformities will imply a loss of profit. It is important to identify possible factors and find a solution. Sometimes these malformations occur due to the use of imbalance diets and, in some cases, prevented adding a supplement to the diet. Tryptophan supplementation was tested using rotifers boosted with liposomes and a dry feed supplemented with tryptophan (see Saavedra et al., 2009b for details on feed formulation). Three different treatments were used: a control consisting of an AA balanced diet, a diet similar to the control but with a supplement of tryptophan and a diet with lysine supplementation. The lysine treatment was used as a second control due to tryptophan low solubility. For that reason the tryptophan diet had a supplement of lysine as well. Liposomes enrichment was done according to Barr and Helland (2007) and the quantity filled in the liposomes are detailed in Saavedra et al. (2009b). The dry feed was given in the form of cross-linked casein-walled capsules as it has lower free amino acid leaching when compared to other kind of inert diets (Yúfera et al., 2002). The control treatment consisted of an AA balanced diet, formulated according to the indispensable amino acid profiles obtained by Saavedra et al. (2006). The dry feed had a percentage of 70% protein, of which 31.1% was casein (minimum required to obtain this type of capsules). All three dry feed diets had the same basic formula but a supplement of lysine and tryptophan was added to the lysine and tryptophan treatments, respectively (see Saavedra et al., 2009b for details on experimental set-up). The effects of the different diets were tested in terms of larval survival and growth and frequency of skeletal deformities.

The results showed that although there were no significant differences in the larval growth and survival rate, larvae fed a supplement of tryptophan had lower dry weight at the end of the experimental trial (Fig. 5).

A high frequency of skeletal deformities was found in this study. When larvae were 15 DAH, the percentage of skeletal deformities was very low and almost no serious malformations were observed. However, when larvae were 25 DAH the frequency of abnormalities increased and kyphosis was found in the trunk vertebrae and scoliosis in the caudal vertebrae (Fig. 6).

The most common vertebral malformation observed in this study, at 15 and 25 DAH, were hypertrophic vertebrae which had a percentage around 70% on 25 DAH larvae (Fig. 6). This malformation affected mainly the trunk and preurostyle vertebrae in 15 DAH larvae and was observed in all regions of the vertebral column in 25 DAH larvae. A small percentage of fusions in the last three vertebrae were observed in all treatments in 15 DAH larvae.

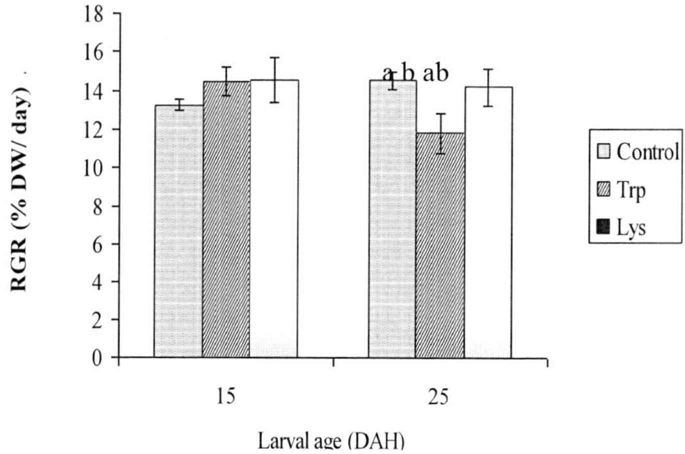

Figure 5. Relative growth rate at 15 and 25 DAH Diplodus sargus larvae fed a different diets. Trp- Tryptophan diet, Lys- Lysine diet. Different letters correspond to significant differences (Data from Saavedra et al., 2009b).

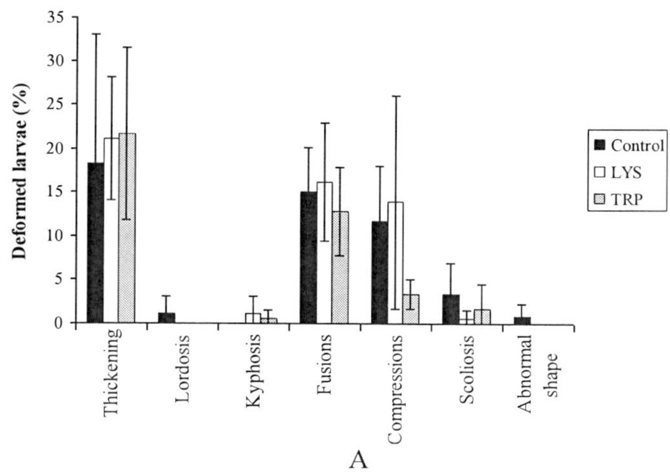

Figure 6A. Continues.

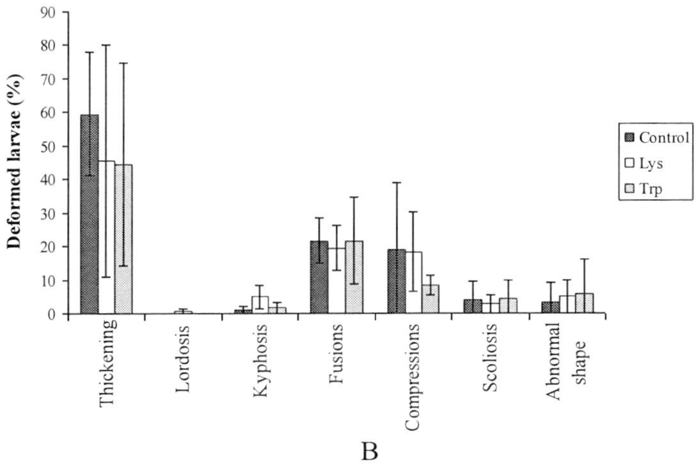

Figure 6. Frequency of skeletal deformities of *Diplodus sargus* larvae fed three different diets: Control, Lysine supplemented diet (Lys) and Tryptophan supplemented diet (Trp) at 15 days after hatched (DAH) (A) and 25 DAH (B) (Data from Saavedra et al., 2010).

Conclusion

Using the *tube feeding* technique was possible to understand the metabolism of tryptophan. This AA had a high evacuation percentage, approximately 70%, compared to the average of all AA which suggests the absorption of tryptophan should be less efficient compared to other AA. Although there is not much information available concerning tryptophan absortion in fish, the fact is that in humans the absorption of this AA seems to be lower compared to other AA (Wahbeh and Christie, 2006). This low absorption of tryptophan is probably related to the specific carrier mediated transport systems involved in the transport across the brush-border membrane. If there are interactions between AA then the AA uptake can be affected, as often transporters have overlapping specificities, therefore decreasing the absorption efficiency of certain AA.

There were no significant differences between AA catabolism and AA retention among treatments. This means that although tryptophan had a low absorption efficiency, it was not more retained when supplemented, suggesting this AA is not limiting growth in *Diplodus sargus* larvae fed rotifers. This was confirmed by the zootechnique trial where larvae fed a diet supplemented with tryptophan did not show an improvement of growth. In fact, larval growth decreased. Such negative effect of a tryptophan supplement was observed by Papoutsoglou et al. (2005a) in rainbow trout juveniles and Papoutsoglou et al. (2005b) in European sea bass juveniles. For both species it was observed a depression of the growth rate and food conversion ratios. Tryptophan may have affected food intake as it is the precursor of serotonin (Schaechter, Wurtman, 1990), which controls appetite. An excessive dose of tryptophan may have increased the serotonin levels and decreased the food intake. Other possible explanation is that tryptophan may have a negative impact on the taste of the microencapsulated diet. This was described for carp to which tryptophan had a deterrent gustatory property and decreased its food consumption (Kasumyan and Morsi, 1996).

During white sea bream larval ontogeny, it was observed a high frequency of skeletal deformities. This incidence did not seem to be improved when larvae were given a supplement of tryptophan. Vertebral thickening or hypertrophic vertebrae, was the most common deformity observed, with a frequency as high as 70%. These skeletal malformations occur more often in earlier larval stages and as larval development continues there is a tendency for the size and shape of the vertebrae to be similar in later phases (Lewis et al., 2004). This was observed when studying *Diplodus sargus* deformity patterns during larval ontogeny. After 25 DAH this type of deformity was no longer observed.

In general, there was an increase in the frequency of skeletal deformities, especially after 15 DAH. Although a higher frequency of deformities were observed, the percentage of serious skeletal deformities such as kyphosis and lordosis were only seen in small percentages.

In conclusion, tryptophan is not limiting growth in *Diplodus sargus* larvae. The supplementation of this AA failed to prevent skeletal deformities in white sea bream larvae and show a negative effect on larval growth, probably either by inhibiting feed consumption or by decreasing the palatability of the diet. The results obtained through a zootechnique trial seem to be consistent with the results from the *tube feeding*.

References

Akiyama, T., Murai, T., Mori, K. 1986. Role of tryptophan metabolites in inhibition of spinal deformity of chum salmon fry caused by tryptophan deficiency. *Bull. Jpn. Soc. Sci. Fish,* 52, 1255-1259.

Baeverfjord, G., Åsgård, T. and Shearer, K.D. 1998. Development and detection of phosphorus deficiency in Atlantic salmon, *Salmo salar* L., parr and post-smolts. *Aquacult. Nutr.* 4, 1-11.

Barr, Y. and Helland, S. 2007. A simple method for mass production of liposome, in particular large liposomes, suitable for delivery of free amino acids to filter feeding zooplankton. *Journal of liposome research* 17, 79-88.

Boglione, C., Gagliardi, G., Scardi, M, Cataudella, S. 2001. Skeletal descriptors and quality assessment in larvae and post-larvae of wild-caught and hatchery-reared gilthead seabream (*Sparus aurata* L. 1758). *Aquaculture,* 192, 1-2.

Cahu, C., Zambonino Infante, J., Tekeuchi, T. 2003. Nutritinal components affecting skeletal development in fish larvae. *Aquaculture* 227, 254-258.

Chatain, B. 1994. Abnormal swimbladder development and lordosis in sea bass (*Dicentrarchus labrax*) and sea bream (*Sparus aurata*). *Aquaculture,* 119, 371-379.

Divanach, P., Boglione, C., Menu, B., Koumoundouros, G., Kentouri, M., Cataudella, S. 1996. Abnormalities in finfish mariculture: an overview of the problem, causes and solutions. In: Chatain, B., Saroglia, M., Sweetman, J., Lavens, P. (Eds.), Seabass and Seabream Culture: Problems and Prospects. European Aquaculture Society, Oostende, Belgium, 45-66.

Divanach, P., Papandroulakis, N, Anastasiadis, P, Koumoundouros, G and Kentouri, M. 1997. Effect of water currents on the development of skeletal deformities in sea bass (*Dicentrarchus labrax* L.) with functional swimbladder during postlarval and nursery phase. *Aquaculture* 156, 145-155.

Favaloro, E. And Mazzola, A. 2000. Meristic character analysis and skeletal anomalies during growth in reared sharpsnout seabream. *Aqua. Int.* 8, 417-430.

Gavaia, P.J., Sarasquete, C. and Cancela, M.L. 2000. Detection of mineralized structures in very early stages of development of marine Teleostei using a modified Alacian blue-Alizarin red double staining technique for bone and cartilage. *Biotech. Histochem.* 75, 79-84.

Halver, J. E. and Shanks, W.E. 1960. Nutition of salmonid fishes VIII. Indispensable amino acids for sockeye salmon. *J. Nutr.* 72, 340-346.

Helland, S., Refstie, S., Espmark, A., Hjelde, K., Baeverfjord, G. 2005. Mineral balance and bone formation in fast-growing Atlantic salmon parr (*Salmo salar*) in response to dissolved metabolic carbon dioxide and restricted dietary phosphorus supply. *Aquaculture*, 250, 364-376.

Helland, S., Denstadli, V., Witten, P.E., Hjelde, K., Storebakken, t., Skrede, A., Asgard, T., Baeverfjord, G. 2006. Hyper dense vertebrae and mineral content in Atlantic salmon (*Salmo salar* L.) fed diets with graded levels of phytic acid. *Aquaculture*, 261, 603-614.

Hseu, J.R., Lu, F.I., Su, H.M., Wang, L.S., Tsai, C.L. and Hwang, P.P. 2003. Effect of exogenous tryptophan on cannibalism, survival and growth in juvenile grouper, *Epinephelus coioides*. *Aquaculture* 218, 251-263.

Kasumyan, A.O., Morsi, A.M. K. 1996. Taste sensitivity of common carp *Cyprinus carpio* to free mino acids and classical taste substances. *J. Ichthyol.* 36, 391-403.

Kloppel, T.M. and Post, G. 1975. Histological alterations in tryptophan-deficient rainbow trout. *J. Nutr.* 105, 861-866.

Koumoundouros, G, Sfakianakis, D.G., Maingot, E., Divanach, P. and Kentouri, M. 2001. Osteological development of the vertebral column and of the fins in *Diplodus sargus* (Teleostei: Perciformes: Sparidae). *Mar. Biol.* 139, 853-862.

Kranenbarg, S., Waarsing, J.H., Muller, M., Weinans, H and Leeuwen, J.L. 2005. Lordotic vertebrae in sea bass (*Dicentrarchus labrax* L.) are adapted to increase loads. *Journal of Biomechanics* 38, 1239-1246.

Lewis, L.M., Lall, S.P. and Witten, P.E. 2004. Morphological description of the early stages of spine and vertebral development in hatchery-reared larval and juvenile Atlantic halibut (*Hippoglossus hippoglossus*). *Aquaculture*, 241, 47-59.

Lim, C. and Lovell, R.T. 1978. Pathology of the vitamin c deficiency syndrome in channel catfish (*Ictalurus punctatus*). *J .of Nutr.* 108, 1137-1146.

Papoustsolgou, S.E., Karakatsouli, N. and Chiras, G. 2005a. Dietary L-tryptophan and tank colour effects on growth performance of rainbow trout (*Oncorhynchus mykiss*) juveniles reared in a recirculating water system. *Aquaculture* 32, 277-284.

Papoutsoglou, S.E., Karakatsouli, N. and Koustas, P. 2005b. Effects of dietary L-tryptophan and lighting conditions on growth performance of European sea bass (*Dicentrarchus labrax*) juveniles reared in a recirculating water system. *J. Appl. Ichthyol.* 21, 520-524.

Rønnestad, I., Rojas-García, C.R., Tonheim, S.K. and Conceição, L.E.C. 2001. In vivo studies of digestion and nutrient assimilation in marine fish larvae. *Aquaculture* 201, 161-175.

Rust, M., Hardy, R.W. and Stickney, R.R. 1993. A new method for force-feeding larval fish. *Aquaculture,* 116, 341-352.

Saavedra, M., Conceição, L.E.C., Pousão-Ferreira, P. and Dinis, M.T. 2006. Amino acid profiles of *Diplodus sargus* (L., 1758) larvae: implications for feed formulation. *Aquaculture,* 261, 587-593.

Saavedra, M., Conceição, L.E.C., Pousão-Ferreira, P. and Dinis, M.T. 2008. Metabolism of tryptophan, methionine and arginine in *Diplodus sargus* larvae fed rotifers: effect of amino acid supplementation. *Amino Acids, 35, 59-64.*

Saavedra, M., Pousão-Ferreira, P., Yúfera, M., Dinis, M.T. and Conceição, L.E.C. 2009a. A balanced amino acid diet improves *Diplodus sargus* larval quality and reduces nitrogen excretion. *Aquac. Nutr.* 15, 517-524.

Saavedra, M., Barr, Y., Pousão-Ferreira, P., Helland, S., Yúfera, M, Dinis, M.T. and Conceição, L.E.C. 2009b. Supplementation of tryptophan and lysine in *Diplodus sargus* larval diet: effects on growth and skeletal deformities. *Aquac. Res.* **40, 1191-1201**.

Saavedra, M., Nicolau, L. and Pousão-Ferreira, P. 2010. Development of deformities at the vertebral column in *Diplodus sargus* (L., 1758) early larval stages. *Aquac. Res.* 41, 1054-1063.

Schaechter JD, Wurtman RJ. 1990. Serotonin release varies with brain tryptophan levels. *Brain Res.* 532, 203–10.

Sfakianakis, D.G., Georgakopoulou, Papadakis, I.E., Divanach, P., Kentouri, M and Koumoundouros, G. 2005. Enviromental determinants of haemal lordosis in European ea bass, Dicentrarchus labrax (Linnaeus, 1758). *Aquaculture,* 254, 54-64.

Yúfera, M. Kolkovski, S., Fernández-Díaz and Dabrowski, K. 2002. Free amino acid leaching from a protein-walled microencapsulated diet for fish larvae. *Aquaculture* 214, 273-287.

Wahbeh, G.T. and Christie, D.L. 2006. Basic aspects of digestion and absorption in: Pediatric gastrointestinal and liver disease. 3rd edition. Edited by Robert Wyllie, MD, Jeffrey S. Hyams, MD and Marsha Kay, pp 9-21.

In: Tryptophan
Editors: B. L. Whitley et al.
ISBN 978-1-62100-445-5

Commentary

Health Benefits from Tryptophan Supplementation in Humans: Is There Sufficient Scientific Evidence?

Daniel Keszthelyi [*]
Department of Internal Medicine, Division of Gastroenterology-Hepatology, Maastricht University Medical Center+, Maastricht, the Netherlands

Abstract

Tryptophan is an essential amino acid and has been used as a dietary supplement for decades because of its alleged health benefits. Apart from its incorporation into body proteins, tryptophan is the precursor for a wide array of metabolites including serotonin, kynurenine, niacin, kynurenic acid and xanthurenic acid, among others.

However, not all metabolites have been identified, and the effects of these substances still remain to be elucidated. Tryptophan is also subject to bacterial degradation in the large intestine, which also gives rise to a

[*] E-mail daniel.keszthelyi@maastrichtuniversity.nl, fax +31 43 387 5006, tel +31 43 388 1982

great number of metabolites. Disturbances in tryptophan metabolism have been associated with a several diseases, including psychiatric conditions and systemic disorders.

Tryptophan also seems to play a key role in immunoregulatory processes through the kynurenine pathway, which has received increased attention over the past few years. Due to the complexity of the metabolic pathways, health effects of nutritional tryptophan supplementation remain controversial. The question arises whether dietary supplementation of tryptophan is in fact able to compensate for any shortage in the human body and if so, what the optimal conditions are for this.

A more complete understanding of tryptophan metabolic pathways will be necessary to apprehend effects of such tryptophan supplementation. This paper aims to provide a brief overview of findings of nutritional tryptophan supplementation in humans and to discuss future perspectives in tryptophan nutritional research.

Introduction

Tryptophan is an essential constituent of the diet. Tryptophan is the precursor of a wide array of metabolites, which are involved in a variety of aspects of human nutrition and metabolism.

Both excessive intake and deficiency of tryptophan are detrimental to health. In the past, tryptophan has been used to treat several conditions, including sleeping disorders, fibromyalgia and depression. However, alleged health benefits of nutritional tryptophan supplementation remain controversial. This commentary analyses the current knowledge of the tryptophan metabolic pathways and summarizes the existing scientific evidence regarding the role of tryptophan supplementation in human nutrition.

Absorption of Tryptophan

L-tryptophan is an essential amino acid, with an estimated dietary requirement of 5 mg/kg per day. It is the limiting amino acid in nearly all protein sources which are of importance for human nutrition, accounting for 1 to 1.5% of total amino acids in typical plant and animal proteins, respectively [1].

Food products containing relatively high tryptophan content are eggs, milk, meat, soybean, potatoes, cereal, broccoli, cauliflower, eggplant, kiwi fruit, plums, bananas, walnuts, fish, seafood and tomatoes. The intestinal absorption of orally ingested tryptophan on the apical membrane of enterocytes is mediated via the B^0AT1 (Solute Carrier 6A19, SLC6A19) epithelial amino acid transport system, which is also responsible for the absorption of all other neutral amino acids and employs a Na^+ co-transport mechanism. With the exception of lysine, all other neutral amino acids have a higher affinity to the transport system than tryptophan. The transporter for tryptophan on the basal membrane of enterocytes is the basolateral aromatic amino acid transporter TAT1 (Slc16a10) protein.

The uptake of tryptophan by peripheral cells, such as tissue macrophages, has not fully been identified yet. Much more is known about tryptophan uptake across the blood brain barrier, which plays a critical role in regulating brain serotonin synthesis. This process is based on competitive transport shared by several large neutral amino acids (LNAAs).

Therefore, increases in the plasma concentration of LNAAs decrease the rate of tryptophan uptake into the brain and hence brain 5-HT synthesis [2]. Very little is known about the mechanism underlying uptake into neurons and other cells in the CNS.

Metabolism of Tryptophan

Tryptophan can enter a number of metabolic pathways: protein synthesis, serotonin pathway, kynurenine pathway and bacterial degradation. Approximately 0,5% of ingested tryptophan is excreted unchanged in urine [3].

Protein Synthesis

Given its limited availability in food, tryptophan is often the rate-limiting amino acid in protein synthesis. There is considerable dispute regarding the extent to which dietary tryptophan is incorporated into protein.

Some authors suggest a majority of 90%[1], others 30% [4], but other sources agree that there is no net new protein synthesis in steady state nitrogen balanced conditions, therefore, the proportion of dietary tryptophan incorporated in protein is minimal [3]. It still remains unresolved how tryptophan is partitioned for protein vs. precursor synthesis.

Serotonin Pathway

About 1-2% of dietary tryptophan is converted to serotonin [3]. Serotonin plays an important role in regulating a number of functions in the human body and serves as the precursor for melatonin synthesis in pinealocytes.

A large bulk of literature exists on the biological role of serotonin in the central nervous system, the gastrointestinal tract and other peripheral organs [5-7]. It is one of the most important signaling molecules within the central nervous system and the gut. Serotonin is synthesized from tryptophan through hydroxylation and decarboxylation. These processes are catalyzed by the tryptophan hydroxylase (TPH) and the aromatic acid decarboxylase (AADC), respectively, the former being the rate-limiting step in the synthesis [5, 6]. Tryptophan hydroxylase (tryptophan 5-monooxygenase) expression is limited to a few specific cells: neurons, pinealocytes, mast cells, mononuclear leukocytes, intestinal enterochromaffin cells and bronchopulmonary neuroendocrine epithelial cells. There are two known isoforms of TPH. TPH1 is localized in enterochromaffin cells and the pineal gland and TPH2 is present in neurons [8]. Both isoforms employ molecular oxygen and the cofactor tetrahydrobiopterin to convert tryptophan to 5-hydroxytryptophan (5-HTP), which is then converted by AADC to serotonin. AADC, which employs vitamin B6 as cofactor, is found in all aminergic cells of the central and peripheral nervous system and is distributed widely in non-neural tissue [9].

Kynurenine Pathway

The kynurenine pathway is the most tryptophan-consuming metabolic pathway. About 95% of the ingested tryptophan enters the kynurenine pathway, which can result in the production of NAD, kynuramines, kynurenic acid, quinolinic acid, picolinic acid but most tryptophan is completely metabolized to yield CO_2 and ATP via the glutarate pathway [1].

The biological functions of the kynurenine pathway are: clearance of excess tryptophan and regulation of plasma tryptophan levels, maintenance of nicotinic acid levels, regulation of CNS function and enhancement of macrophage defense function. Besides the serotonin pathway, which has well been characterized over the past decades, the kynurenine pathway seems to emerge as an important regulator of several biological functions in the human body [10].

Entering the kynurenine pathway, tryptophan is first oxidized by tryptophan 2,3-dioxygenase (TDO), which is almost entirely localized at hepatic cells. TDO is the rate limiting enzyme for kynurenine synthesis in the periphery. TDO expression and activity can be induced four- to tenfold by tryptophan loading within a period of a few hours [3]. The principal branch of the kynurenine pathway generates quinolinic acid and nicotinamide, whereas the side chains generate kynurenic acid and xanthurenic acid. Several biological features of kynurenine metabolites have been described. Most attention has been given to the imbalance in neurotoxic and neuroprotective properties of these compounds, which have been associated with several CNS pathologies. Quinolinic acid is considered to be an excitotoxic N-methyl D-aspartate (NMDA) receptor agonist, whereas kynurenic acid is a neuroprotective NMDA antagonist and an $\alpha 7$ nicotinic cholinergic agonist [11]. In mononuclear cells, including tissue macrophages, quinolinic acid is the main end product of the kynurenine pathway and plays a role in immunoregulatory processes [12].

The kynurenine pathway also provides the precursors for the dietary supplement niacin, a collective term for nicotinamide and nicotinic acid. Under normal conditions, most of the tryptophan that enters the oxidative pathway is converted to CO_2 and water in the glutarate pathway. Only if this branch of the pathway is saturated, NAD becomes a major product of metabolism [1]. Although metabolites of the glutarate pathway are present in many tissues, including the intestine, NAD synthesis is only possible in the liver, because this is the only organ that possesses all the necessary enzymes [12].

Another product of the kynurenine pathway is picolinic acid. Picolinic acid is only produced when the flux of metabolites through the glutarate pathway is high and enzymes of the glutarate pathway are saturated [1]. Picolinic acid acts as a chelating agent of elements such as chromium, zinc, manganese, copper, iron, and molybdenum in the human body. It forms a complex with zinc that may facilitate the passage of zinc through the gastrointestinal wall and into the circulatory system [3]. Several of the enzymes of the kynurenine pathway use vitamin B6 as cofactor and a key feature of the enzyme kynureninase is its exceptionally high sensitivity to pyridoxine deficiency. Lack of vitamin B6 leads to a large increase in xanthurenic acid excretion. This has been used for decades as a diagnostic test for vitamin B6 deficiency [3]. Vitamin B6 deficiency also compromises serotonin synthesis, and hence can lead to competition between the two pathways for the co-factor.

Besides TDO, another enzyme initializing the kynurenine pathway is indoleamine 2,3-dioxygenase (IDO). IDO is widely distributed in peripheral tissues. The human intestine contains a relatively large amount of IDO [13]. While TDO exclusively accepts tryptophan as substrate, IDO has a broader specificity and can also take 5-HTP, 5-HT and tryptamin [13]. The expression of IDO increases in response to infection and inflammation, with interferon-γ being the strongest stimulator. Mononuclear cells that synthesize IDO reduce extracellular tryptophan concentration so that adjacent T-cells, which depend on tryptophan from the extracellular environment, are unable to activate and proliferate upon encountering antigens. Therefore, IDO might play a role in preventing the initiation of autoimmune disease by enforcing T-cell tolerance through suppressing their proliferation [12]. Hence, high local expression of IDO by mononuclear cells may represent an anti-inflammatory and immunosuppressive mechanism tempting to counterbalance tissue damage [14]. This mechanism could be involved in intestinal pathophysiology, since IDO expression is markedly induced in lesional colonic biopsies of inflammatory bowel disease (IBD) patients [14] and increased IDO activity has been observed in patients with celiac disease [15] and diverticulitis [16]. A similar IDO-based intrinsic immunoescape mechanism is probably employed by colon tumor cells [17].

Besides through the regulatory effect of IDO on T-cells and immune function, inflammatory responses related to the kynurenine pathway can also be based on a sensitive balance between the pro-inflammatory, excitotoxic quinolinic acid and the anti-inflammatory, neuroprotective kynurenic acid [18]. This balance could have profound influence on the excitability of neurons, and can contribute to pathologies of the CNS and the periphery. Increased levels of the kynurenine pathway metabolites have been associated with diseased states such as depression [11], schizophrenia [19] and irritable bowel syndrome [20].

Other products of the IDO and formamidases are kynuramin derivates. Formation of kynuramines has been described in various tissues, including the intestine and appears to be directly proportional to tryptophan concentrations. Kynuramines may be important as endogenous agonists or antagonists of 5-HT receptors in smooth muscle. Marked non-selective serotonergic agonist properties of 5-hydroxykynuramine at multiple 5-HT receptors were demonstrated in rat ileum. 5-hydroxykynuramine is formed from tryptophan to much lesser extent *in vivo* than 5-HT, but pathological conditions or situations in which tryptophan concentrations are increased may lead to an overproduction of kynuramines. Besides its effect on smooth muscle, 5-

hydroxykynureamine is a potent inhibitor of the action of serotonin in promoting the aggregation of platelets. This may provide a measure of regulation in cases of over-synthesis of serotonin, not only as an alternative catabolic pathway of the amine, but also to inhibit one of its biological actions [21].

Evidence also suggests that the kynurenine pathway itself can be influenced by nutrition [12]. Kynurenic acid has been proposed to have a positive effect stroke, epilepsy, neurodegenerative disorders [19] and in conditions accompanied with intestinal hypermotility [18]. It is present in certain sources of food, such as honeybee products [22]. Therefore, kynurenic acid and its derivates may potentially be used as a therapeutic agent, possibly through nutritional intervention.

Bacterial Degradation Products

Approximately 4-6% of tryptophan undergoes bacterial degradation yielding indole, indican, and indole acid derivates [1, 3].

Indican. The main bacterial breakdown product of tryptophan is indole. The wide range of bacterial species capable of producing indole include *E. coli*, *Proteus vulgaris*, *Paracolobactrum coliforme*, *Achromobacter liquefaciens* and *Bacteriodes* spp. The formation of indole is catabolyzed by the enzyme tryptophanase, which is inducible by tryptophan and repressible by glucose in most bacteria [23]. By-products of this conversion are pyruvate, which can be used in fermentation or respiration reactions, and ammonia, which can have potentially toxic effects on the intestinal epithelium. High protein diets are therefore able to induce bacterial tryptophanase activity, which in turn results in overproduction of indole and other compounds that can thereby reach toxic concentrations in the colon [23].

After absorption, indole is oxidized to indoxyl, conjugated with sulphate and excreted as urinary indican (also known as indoxyl-sulphate). In normal individuals only small proportions of dietary tryptophan reach the colon because of nearly complete absorption in the small intestine. Approximately 3% of dietary tryptophan is excreted as urinary indican [3].

Indican is also known to be a nephrotoxin that accumulates in the blood of patients suffering from chronic kidney failure. Because tryptophanase activity derives from only a subset of enteric bacteria, non-indole-producing bacteria, such as various *Bifidobacterium* species, have been administered as a test probiotic to dialysis patients to decrease their plasma levels of indoxyl sulfate [24].

Indolic acid derivates. A smaller quantity of tryptophan is converted by bacterial action to indolic acid derivates: indolyl-3-acetic acid, indolyl-acetyl-glutamine, indolyl-propionic acid, indolyl-lactic acid, indolyl-acrylic acid and indolyl-acryloyl-glycine. Intestinal microorganisms, including *Bacteriodes*, *Clostridia* and *E.coli*, catalyze tryptophan to tryptamin and indolyl-pyruvic acid, which are then converted to indolyl-3-acetic acid, indolyl propionic acid and indole lactic acid [23]. Indolyl acetic acid can be conjugated with glutamine in the liver to yield indolylacetyl glutamine. Indolylpropionic acid can be further converted in the liver or kidney into indolyl acrylic acid (IAcrA) and conjugated with glycine to produce indolylacryloyl glycine (IAcrGly). Some evidence suggests that IAcrA can also be produced in the absence of intestinal microorganisms, although there is no direct evidence for enzymatic or non-enzymatic processes [25]. Nutritional intervention such as tryptophan loading did not influence urinary IAcrGly levels, but complete elimination of tryptophan from the diet resulted in a marked decrease of IAcrGly in urine, similarly to parenteral alimentation [25].

The biological role of these compounds still needs to be investigated. Increased and prolonged excretion of urinary indols (e.g. indican, indolyl-3-acetic acid, indolyl-3-acetyl-glutamine, indolyl-lactic acid, indolyl-acryloyl-glycine) has been observed in a number of diseased states including Hartnup disorder, celiac disease and other malabsorptive states [26]. It is assumed that this is due to excessive tryptophan overload in the colon possibly with coexistent alteration in gut microbiota, which leads to increased production of bacterial degradation products [26].

Some bacterial products are toxic to other microbiota, and this provides competitive advantage for the producers. Some indolic compounds are known to have bacteriostatic effect on Gram negative enterobacteria, especially within the genera *Salmonella* and *Shigella* [23]. The increased urinary excretion of indolic compounds reflects variations in gut microbiota composition in relation to nutritional competition. For instance, indolyl acetic acid has been reported to inhibit the growth and survival of *Lactobacilli*, and specifically *L. paracasei* [27]. Also, indolyl propionic acid has been shown to be a powerful antioxidant, and is currently being investigated as a possible treatment for Alzheimer's disease [24].

The biological effects of bacterial degradation products of tryptophan need to be further clarified as well. An overload of dietary proteins or changes in intestinal microbiota can lead to overproduction of these products and subsequently lead to changes in intestinal physiology with potentially toxic consequences. Interest in the impact of gut microbial activity on human health is expanding rapidly, and a better identification of the host-microbiota interactions can also lead to developing new approaches in disease treatment. Probiotic supplementation, for instance, can aim to replace or reduce the number of potentially harmful proteolytic *E.coli* and *Clostridia* producing toxic tryptophan breakdown products by enriching populations of gut microbiota that have more advantageous metabolic activity. Future activities will be directed to influence the gut microbiota in a targeted way, ideally by enhancing beneficial effects and minimizing adverse effects. However, before we are able to do so, further work is required to understand in more detail the processes underlying the bacterial conversion of tryptophan and other dietary components.

Nutritional Supplementation of Tryptophan

The question arises to which extent tryptophan and serotonin metabolism can be influenced by nutrition and how this can potentially exert beneficial health effects. In-depth understanding of pathways involved in tryptophan and serotonin metabolism is therefore crucial. Furthermore, the interrelationship between metabolic pathways is of profound pharmacological and physiological importance, in that changes in one pathway might have secondary effect on the others.

Nutritional intervention has been employed since decades in investigating the serotonergic system, which has been proven to be involved in the pathogenesis of several psychiatric disorders, mainly depression. Furthermore, serotonin also has a pivotal role in regulating appetite, satiety and food intake [28].

Historically, tryptophan has been used as an anti-depressive and sleep-inducing substance through increasing serotonin synthesis [29]. Similar effects have been observed in case of supplementation with a tryptophan-rich protein, alpha-lactalbumine [30] and nutritionally-sourced tryptophan [31].

These effects are theoretically based on the Wurtman hypothesis [32]. Accordingly, dietary intake of tryptophan, especially in combination with carbohydrates, is able to increase serotonin synthesis in the CNS. However, a Cochrane review on the has concluded that there is insufficient evidence to support a role for tryptophan in the treatment of depression and that further studies were needed to assess efficacy and safety [33]. Availability of alternative antidepressants that are both safe and effective also hamper extensive research on these effects of tryptophan. Also, a very recent scientific opinion paper from the European Food Safety Authority has found no beneficial effect of L-tryptophan on the maintenance of normal sleep, the enhancement of mood, the contribution to normal cognitive function, or the maintenance of normal body weight for the general population [34]. In addition, a large epidemiological study of almost 30 thousand men found no association between dietary tryptophan intake and self-reported depressed mood or hospital admission for depressive disorder [35].

Theories of alleged beneficial effects of nutritional tryptophan supplementation face further practical and theoretical issues. Starting in November 1989, a disease termed eosinophilia myalgia syndrome (EMS) was reported, associated with L-tryptophan sold over-the-counter in the USA. EMS is a complex disorder with inflammatory and autoimmune components that affect the skin, fascia, muscle, nerves, blood vessels, lungs and heart. More than 1500 cases and 30 deaths have been reported. It has been suggested that these cases were caused by impurities in some batches of L-tryptophan from a specific manufacturer. Several studies have been conducted in the past decade to investigate the exact nature of these cases. Methodological limitations in epidemiologic studies and the inability to identify the exact pathogenic components of the implicated batches of tryptophan have however hampered the understanding of disease pathogenesis. Although a recent study has identified a number of immunogenetic factors contributing to EMS associated with L-tryptophan intake [36], uncertainty still remains regarding the exact mechanisms involved. Therefore, until such details are elucidated, caution should still be observed when advising to take L-tryptophan as a food supplement.

Furthermore, from a theoretic point of view, tryptophan loading is known to increase catabolism along the kynurenine pathway as tryptophan is able to induce the enzymes that initiate this pathway. Also, all access of tryptophan is cleared via this metabolic branch under normal circumstances. This implies that dietary tryptophan loading results in increased oxidative breakdown of tryptophan along the kynurenine pathway without necessarily leading to an increased serotonin synthesis. Even more so, an overload of tryptophan can potentially be harmful as it can produce kynurenine metabolites such as quinolinic acid that are known to be neurotoxic. A study by Forrest et al. has shown that 6 g oral load of tryptophan in healthy adults significantly increased lipid peroxidation after 5 and 7 h compared to baseline [37]. The oxidative stress was suggested to result from the generation of quinolinic acid, 3-hydroxykynurenine, and 3-hydroxyanthranilic acid, all of which are known to have the ability to generate free radicals. A more comprehensive understanding of the tryptophan metabolic pathways will potentially enable us to selectively influence levels of certain metabolites, which can have beneficial health effects.

CONCLUSION

General media reports regarding the beneficial health effects of tryptophan are often hyperbolic, misleading and ambiguous, with much of the advice often inconsistent even with basic biochemistry. Effects of dietary tryptophan loading or supplementation on different tryptophan metabolic pathways and hence their biological effects are still not fully understood. Given the current knowledge, evidence is insufficient to support alleged health effects of dietary tryptophan supplementation in the general population. In clinical practice, availability of alternative treatment options for depression, insomnia and other related conditions do not support the use of tryptophan as a therapeutic entity either. However, this does not detract from the importance of a nutritious diet for individuals with mood disorders as well as the general population, for overall health. Certainly, further studies on dietary tryptophan supplementation and tryptophan metabolic pathways, especially the kynurenine pathway, will contribute to our overall scientific knowledge. Future work should investigate pathways linked to tryptophan in order to obtain a more complete picture of the net effects of tryptophan loading or supplementation. This can indeed open new horizons in exploring possible therapeutic applications for tryptophan nutritional interventions, in particular by specifically increasing or decreasing production of certain metabolites within the tryptophan metabolic pathways resulting in beneficial health effects.

REFERENCES

[1] Peters, J.C., Tryptophan nutrition and metabolism: an overview. *Adv. Exp. Med. Biol*, 1991. 294: p. 345-58.

[2] Fernstrom, J.D., Branched-chain amino acids and brain function. *J. Nutr*, 2005. 135(6 Suppl): p. 1539S-46S.

[3] Bender, D.A., Biochemistry of tryptophan in health and disease. *Mol. Aspects Med*, 1983. 6(2): p. 101-97.

[4] Allegri, G., et al., Enzyme activities of tryptophan metabolism along the kynurenine pathway in various species of animals. *Farmaco*, 2003. 58(9): p. 829-36.

[5] Gershon, M.D. and J. Tack, The serotonin signaling system: from basic understanding to drug development for functional GI disorders. *Gastroenterology*, 2007. 132(1): p. 397-414.

[6] Lesurtel, M., et al., Role of serotonin in the hepato-gastroIntestinal tract: an old molecule for new perspectives. *Cell Mol. Life Sci,* 2008. 65(6): p. 940-52.

[7] Berger, M., J.A. Gray, and B.L. Roth, The expanded biology of serotonin. *Annu. Rev. Med*, 2009. 60: p. 355-66.

[8] Walther, D.J., et al., Synthesis of serotonin by a second tryptophan hydroxylase isoform. *Science*, 2003. 299(5603): p. 76.

[9] Chang, Y.T., G. Mues, and K. Hyland, Alternative splicing in the coding region of human aromatic L-amino acid decarboxylase mRNA. *Neurosci. Lett*, 1996. 202(3): p. 157-60.

[10] Keszthelyi, D., F.J. Troost, and A.A. Masclee, Understanding the role of tryptophan and serotonin metabolism in gastrointestinal function. *Neurogastroenterol. Motil*, 2009. 21(12): p. 1239-49.

[11] Stone, T.W. and L.G. Darlington, Endogenous kynurenines as targets for drug discovery and development. *Nat. Rev. Drug Discov*, 2002. 1(8): p. 609-20.

[12] Moffett, J.R. and M.A. Namboodiri, Tryptophan and the immune response. *Immunol. Cell Biol*, 2003. 81(4): p. 247-65.

[13] Yamazaki, F., et al., Human indolylamine 2,3-dioxygenase. Its tissue distribution, and characterization of the placental enzyme. *Biochem. J,* 1985. 230(3): p. 635-8.

[14] Wolf, A.M., et al., Overexpression of indoleamine 2,3-dioxygenase in human inflammatory bowel disease. *Clin. Immunol*, 2004. 113(1): p. 47-55.

[15] Torres, M.I., et al., Tryptophan metabolism and indoleamine 2,3-dioxygenase expression in coeliac disease. *Clin Exp Immunol*, 2007. 148(3): p. 419-24.

[16] Ferdinande, L., et al., Inflamed intestinal mucosa features a specific epithelial expression pattern of indoleamine 2,3-dioxygenase. *Int. J. Immunopathol. Pharmacol*, 2008. 21(2): p. 289-95.

[17] Uyttenhove, C., et al., Evidence for a tumoral immune resistance mechanism based on tryptophan degradation by indoleamine 2,3-dioxygenase. *Nat. Med*, 2003. 9(10): p. 1269-74.

[18] Kaszaki, J., et al., Kynurenic acid inhibits intestinal hypermotility and xanthine oxidase activity during experimental colon obstruction in dogs. *Neurogastroenterol. Motil*, 2008. 20(1): p. 53-62.

[19] Stone, T.W., Kynurenines in the CNS: from endogenous obscurity to therapeutic importance. *Prog. Neurobiol*, 2001. 64(2): p. 185-218.

[20] Forrest, C.M., et al., Levels of purine, kynurenine and lipid peroxidation products in patients with inflammatory bowel disease. *Adv. Exp. Med. Biol*, 2003. 527: p. 395-400.

[21] Hayaishi, O., Properties and function of indoleamine 2,3-dioxygenase. *J. Biochem*, 1976. 79(4): p. 13P-21P.

[22] Turski, M.P., et al., Presence of kynurenic acid in food and honeybee products. *Amino Acids*, 2008.

[23] Smith, E.A. and G.T. Macfarlane, Formation of Phenolic and Indolic Compounds by Anaerobic Bacteria in the Human Large Intestine. *Microb. Ecol*, 1997. 33(3): p. 180-8.

[24] Wikoff, W.R., et al., Metabolomics analysis reveals large effects of gut microflora on mammalian blood metabolites. *Proc. Natl. Acad. Sci. U S A,* 2009. 106(10): p. 3698-703.

[25] Marklova, E., Where does indolylacrylic acid come from? *Amino Acids*, 1999. 17(4): p. 401-13.

[26] Haverback, B.J., B. Dyce, and H.V. Thomas, Indole metabolism in the malabsorption syndrome. *N. Engl. J. Med*, 1960. 262: p. 754-7.

[27] Nowak, A. and Z. Libudzisz, Influence of phenol, p-cresol and indole on growth and survival of intestinal lactic acid bacteria. *Anaerobe*, 2006. 12(2): p. 80-4.

[28] Gruninger, T.R., et al., Molecular signaling involved in regulating feeding and other motivated behaviors. *Mol. Neurobiol*, 2007. 35(1): p. 1-20.

[29] Agazzi, A., et al., Review of the implications of dietary tryptophan intake in patients with irritable bowel syndrome and psychiatric disorders. *Dig. Liver Dis*, 2003. 35(8): p. 590-5.

[30] Markus, C.R., et al., Evening intake of alpha-lactalbumin increases plasma tryptophan availability and improves morning alertness and brain measures of attention. *Am. J. Clin. Nutr*, 2005. 81(5): p. 1026-33.

[31] Attenburrow, M.J., et al., The effect of a nutritional source of tryptophan on dieting-induced changes in brain 5-HT function. *Psychol. Med*, 2003. 33(8): p. 1381-6.

[32] Wurtman, R.J. and J.J. Wurtman, Carbohydrates and depression. *Sci. Am*, 1989. 260(1): p. 68-75.

[33] Shaw, K., J. Turner, and C. Del Mar, Tryptophan and 5-hydroxytryptophan for depression. *Cochrane Database Syst. Rev*, 2001(3): p. CD003198.

[34] EFSA Panel on Dietetic Products, N.a.A.N., Scientific Opinion on the susbtantiation of health claims related to L-tryptophan and maintanace of normal sleep (ID 596, 1671), enhancement o mood (ID 596), contribution to normal cognitive function (ID 596) and contribution to the maintenace or achievement of a normal body weight (ID 604) pursuant to Article 13(1) of Regulation (EC) No 1924/2006. *EFSA Journal*, 2011. 9(4): p. 2073.

[35] Hakkarainen, R., et al., Association of dietary amino acids with low mood. *Depress Anxiety*, 2003. 18(2): p. 89-94.

[36] Okada, S., et al., Immunogenetic risk and protective factors for the development of L-tryptophan-associated eosinophilia-myalgia syndrome and associated symptoms. *Arthritis Rheum*, 2009. 61(10): p. 1305-11.

[37] Forrest, C.M., et al., Tryptophan loading induces oxidative stress. *Free Radic Res*, 2004. 38(11): p. 1167-71.

Index

C

D

I

K

L

M

N

O

S

T

U

V

W

X

Y

Z